VISUAL COMMUNICATION

This smart and engaging book introduces readers to some of the most pressing debates in visual communications studies. Effective in making often-opaque theories accessible, Aiello and Parry situate these debates in relation to a diverse toolkit of current research methods. Eighteen thickly-contextualized case studies skilfully illustrate the various steps students need to design their own projects.

Wendy Kozol, Oberlin College

How can we understand images in media culture? Writing with clarity, insight and flair, Aiello and Parry show us that while there is no simple answer, there are many good analytical paths to pursue, demonstrating their value across no less than eighteen case studies – from political memes to photojournalism to Hollywood movie trailers and commercial imagery. Focusing on how visual communication is entangled with identities, politics and commodities, this book is not only an exemplary introduction to visual communication research: it is a significant and timely guide to the powers and properties of contemporary images.

Paul Frosh, Hebrew University of Jerusalem

Organised around the themes of identities, politics and commodification, this book offers multiple conceptual insights into how images are created, circulated, seen, sold, modified and destroyed. Its themes and arguments are grounded in a series of detailed and clearly written examples, which also explore the methodological implications of approaching images as forms of visual communication. All this adds up to a must-read for anyone interested in contemporary visual culture.

Gillian Rose, University of Oxford

VISUAL COMMUNICATION
Understanding Images in Media Culture

Giorgia Aiello
& Katy Parry

Los Angeles | London | New Delhi
Singapore | Washington DC | Melbourne

Los Angeles | London | New Delhi
Singapore | Washington DC | Melbourne

SAGE Publications Ltd
1 Oliver's Yard
55 City Road
London EC1Y 1SP

SAGE Publications Inc.
2455 Teller Road
Thousand Oaks, California 91320

SAGE Publications India Pvt Ltd
B 1/I 1 Mohan Cooperative Industrial Area
Mathura Road
New Delhi 110 044

SAGE Publications Asia-Pacific Pte Ltd
3 Church Street
#10-04 Samsung Hub
Singapore 049483

Editor: Michael Ainsley
Editorial assistant: Amber Turner-Flanders
Production editor: Imogen Roome
Copyeditor: Diana Chambers
Proofreader: Joanna North
Marketing manager: Lucia Sweet
Cover design: Francis Kenney
Typeset by: C&M Digitals (P) Ltd, Chennai, India
Printed in the UK by Bell and Bain Ltd, Glasgow

© Giorgia Aiello and Katy Parry 2020

First published 2020

Apart from any fair dealing for the purposes of research or private study, or criticism or review, as permitted under the Copyright, Designs and Patents Act, 1988, this publication may be reproduced, stored or transmitted in any form, or by any means, only with the prior permission in writing of the publishers, or in the case of reprographic reproduction, in accordance with the terms of licences issued by the Copyright Licensing Agency. Enquiries concerning reproduction outside those terms should be sent to the publishers.

Library of Congress Control Number: 2019940965

British Library Cataloguing in Publication data

A catalogue record for this book is available from the British Library

ISBN 978-1-4129-6223-0
ISBN 978-1-4129-6224-7 (pbk)

At SAGE we take sustainability seriously. Most of our products are printed in the UK using responsibly sourced papers and boards. When we print overseas we ensure sustainable papers are used as measured by the PREPS grading system. We undertake an annual audit to monitor our sustainability.

CONTENTS

About the authors — vii
Acknowledgements — ix
Preface — xi

1 Introduction — 1
2 Understanding images in media culture: methodological considerations — 17

PART I IDENTITIES

3 Envisioning the self in digital media — 37
4 Communicating visions of collective identity — 61
5 Ways of seeing difference beyond stereotypes — 85

PART II POLITICS

6 Images of politicians in the public sphere — 109
7 The visual spectacles of protest and activism — 135
8 Picturing international conflict and war — 159

PART III COMMODITIES

9 The visual attractions of advertising and promotional culture — 185
10 Visualizing lifestyles as commodities — 209
11 Brands as visual experiences — 233

12 Conclusion — 257

References — 267
Index — 289

ABOUT THE AUTHORS

Giorgia Aiello is an Associate Professor in Media and Communication at the University of Leeds. Her research focuses on the politics and potentials of visual, multimodal and material communication. Giorgia is interested in how aesthetics shape and are shaped by political, economic and cultural agendas. Her work aims to uncover how identities are formed, how both difference and diversity are negotiated, and how inequalities are maintained or overcome through media and communication. She is a co-editor of *Communicating the City: Meanings, Practices, Interactions* (Peter Lang, 2017), with Matteo Tarantino and Kate Oakley, and the author of many journal articles, book chapters and short publications about branding, photography, data visualization, political imagery and cities.

Katy Parry is an Associate Professor in Media and Communication at the University of Leeds. Her work focuses on visual politics and activism, images of war and representations of contemporary soldiering. She is a co-author of *Political Culture and Media Genre: Beyond the News* (Palgrave Macmillan, 2012) with Kay Richardson and John Corner, and a co-editor of *Can the Media Serve Democracy? Essays in Honour of Jay G. Blumler* (Palgrave Macmillan, 2015), with Stephen Coleman and Giles Moss. Her current research interests include the visual expressions of political solidarity shared in response to the murder of Jo Cox MP during the Brexit referendum campaign, and the representational politics of the war veteran in contemporary media.

ACKNOWLEDGEMENTS

Like many academic book projects, this one has had a long gestation period. We are very grateful to Mila Steele who first commissioned the book for SAGE, and we are indebted to the editorial SAGE team, particularly Michael Ainsley and John Nightingale, for their patience and support throughout. We also thank the anonymous reviewers who provided very helpful advice on key draft chapters.

We are lucky to be surrounded with supportive colleagues in the School of Media and Communication at the University of Leeds, and have also benefited from insightful discussions with friends and colleagues from all over the world at various conferences and research events. There are too many to name here, but we would like to specifically acknowledge those who have supported and advised us as mentors over the last several years: Stephen Coleman, John Corner, Myria Georgiou, David Hesmondhalgh, Helen Kennedy, Nancy Thumim, Crispin Thurlow and Katrin Voltmer.

We are very grateful to the COMM2125 Visual Communication student classes of 2016–17 and 2017–18 at the University of Leeds who participated in our surveys and focus groups for the case studies in Chapters 8 and 11. We thank those who have granted us permission to reprint images, including Peter DiCampo for Everyday Africa, Getty Images, Reuters, Shutterstock and Jeanette Martin.

We would also like to acknowledge our own writing partnership, sustained with laughter, hugs, Marmite toast and Italian coffee. This is a co-authored book that brings together our shared fascination with all things visual, but inevitably we have taken the lead on writing chapters that chimed with our strengths and research interests: for Giorgia, this was Chapters 2, 3, 4, 10 and 11; and for Katy, Chapters 5, 6, 7, 8 and 9.

Finally, we thank our families and partners, mainly for being there for us and for talking about subjects unrelated to visual communication when needed. Giorgia is thankful to her mother, Chiara, and her late father, Enzo, for their unconditional love and encouragement throughout her studies and career, even though both took her far away from home. She would like to extend a special thanks to her partner Chris Anderson, who read and commented on several draft chapters, was a great sport about the many evenings and weekends that were spent working on this book, and was and continues to be just generally brilliant. Katy thanks the marvellous Parry clan – Mum (Jane) and Dad (Chris), Dave, Liza, Muka, Ria, Melissa, Joe, Ben, Wilco and Edwin – for enduring her ardent views, usually expressed with the help of (too much) wine. She especially thanks her partner Paul for reading draft chapters, but really for everything, and Frank for the all-important dog walks. She feels very lucky to be part of such a loving, creative, anarchic family.

PREFACE

The eyes look up to the right side of the frame, a 'three-quarter' gaze associated with an ennobling vision of the future. Only the lips and the mascaraed, slightly bloodshot brown eyes are visible; other facial features remain hidden beneath the red knitted ski mask. But the eyes and lips convey femininity within the roughly cut and fraying holes of the mask. Caught in close-up, the red hues of the woolly mask dominate the image. The shape of the face is highlighted in the ribbed texture of the material and the lighting that softens the colouring from a dark red at the chin, to pinkish tones on the covered forehead.

Who is the woman photographed in the mask on the front cover of this book? For those familiar with the Russian punk protest group, Pussy Riot, the homemade brightly coloured balaclava is an instantly recognizable symbol of their visually striking activism. The photograph was taken by Ryan Pierse during the 2014 Winter Olympic Games in Sochi, Russia, where Pussy Riot had staged a protest. Two prominent members of the group, Nadya Tolokonnikova and Maria Alyokhina, had been released from jail just a few months earlier in December 2013. They had served most of their two-year sentences for performing a 'punk prayer' in Moscow's main cathedral, with lyrics like 'Virgin Mary, Mother of God, banish Putin, banish Putin,/Virgin Mary, Mother of God, banish him, we pray thee!' and 'Be a feminist, we pray thee./Bless our festering bastard-boss.'

Why did we decide on this photograph? Choosing a cover design for a book about visual communication is both challenging and anxiety-inducing, with no single image able to capture the diversity of visual media we cover in the book. Pussy Riot feature as a case study in our chapter about visually oriented political activism (Chapter 7), but we thought that this bold image conveyed something further about the way in which media images are both evidential and ambiguous, revealing and masking, easily read and enigmatic.

The photograph speaks to the three overlapping thematic strands of this book – identities, politics, commodities. Our interest in **identities** looks to the role played by visual images in the shaping and maintaining of identities. The DIY aesthetic of the frayed ski mask serves as an identifying symbol of Pussy Riot. Indeed, as we explain in our case study, Pussy Riot see the costumes as performative: 'You put on the mask and you become Pussy Riot.' This is claimed as part of their democratizing ethos; membership is open to anyone with the right punk attitude and bravery to challenge the Putin government through actions designed to attract global media attention. However, the mask also hides the wearer's personal identity, in the style of a mischievous anarchist superhero designed to unsettle those in power while signalling solidarity with dissenters.

The cover image therefore also speaks to our interest in **politics**, and specifically in how protesters harness the rhetorical power of the visual to fight the powerful. To simply bemoan the encroaching theatrical aspects of political life is also to deny that politics has long embraced drama and performance. Visual images are a crucial element in expressions of the need for radical change, of imagining other ways of being, and exposing the absurdities of the status quo. Symbols of solidarity are shared via media technologies across the world, and in some cases they are formed around icons of injustice to expose abuses of power.

The masks have been referred to as the 'trademark' look of Pussy Riot, and the choice of words hints at the potential for the Pussy Riot look to become **commoditized**. Pussy Riot members are aware of their potential to be cast as trivial or ridiculous, deliberately playing with their media image. The idea that Pussy Riot are a 'machine placed inside the media', as suggested by a source in our case study, hints at a purposeful media strategy that may be more easily associated with a viral ad campaign than a political protest. The ethos and aesthetic of handmade or recycled clothing is supposed to be set against the cultural homogeneity of shop-bought items. With colourful Pussy Riot-themed balaclavas for sale online, however, their creative style cannot escape being subsumed into the market.

There is a second way we can reflect on the image as a commodity, and that is our own use in this book. Building on our idea to use the case study of Pussy Riot, our publishers suggested this image to us, after having their design team locate it in the online database of Getty Images, the world-leading corporate image bank. As a commodity in its own right, Ryan Pierse's editorial photograph of Pussy Riot has several advantages over in-house or commissioned imagery. It is unique but can be easily licensed for a relatively small fee, and it is also both high quality and ready to use, thus eschewing the costs and delays that would be otherwise associated with the production of a cover image from the bottom-up. From the publishers' perspective, this is also a bold and colourful photograph, one that stands out on a bookshelf and as an online thumbnail image. In the back-and-forth process of deciding on the exact design, then, we also had to consider the book as a commodity. What do we want the cover to say about our work and who we are?

Locating the photograph in an image bank also speaks to other themes that we discuss in this book. How are photographs (and their meanings) recontextualized in new media spaces? What role do dominant photographic agencies, global image banks and overall powerful image-makers play in this process? How do the technological affordances of image collections, including the use of tags and keywords, promote certain representational types?

Over the pages of this book, we present discussion and analysis of the varied ways in which elements of visual communication play a vital role in the meaning-making practices encountered across a range of media. We show how examining the visual recipes and characterizations of media images helps us to detect the values underpinning such representations, while explaining their role in further shaping the dynamics of contemporary society, culture and politics.

# 1 INTRODUCTION

'We are now bombarded with images.'

'We are living in a more visually led world.'

'The internet has democratized image production and sharing.'

Students, journalists and scholars all too often make proclamations like the ones above in order to underscore the importance of paying close attention to media images. We have all read (and in some cases also written) endless essays, articles and books about visual communication that open with general if not sweeping statements of this kind. These are, of course, well-intentioned attempts to capture not only the abundance but also the vibrancy of media images in our daily encounters. In this book, we recognize the centrality of images in meaning-making processes as we continue to marvel at their unprecedented proliferation and pervasiveness. Our purpose is to unpack generalized assertions and examine how images communicate in distinctive manners across varied contexts – and through a range of media formats, genres and platforms.

In recent years, visual communication has become a regular feature of everyday activities for a wide variety of individuals, groups and institutions. Not only are visual images the bread and butter of how ordinary people and media professionals communicate, but the work of corporations, governments and activists is increasingly dependent on their ability to craft distinctive graphics and iconic imagery. The existence of multiple technological platforms for the production, distribution and consumption of media content has led to the proliferation of images in multifaceted communication environments such as journalism, branding and social networking. And while the visual has long been a substantial mode of communication, it has become all the more urgent to examine the implications of images in the light of the digital, global and multimodal characteristics of present-day media.

Although there is no dearth of books on visual images, this volume fills a gap in the literature. Most books on visual communication focus on particular theories and issues, or introduce the reader to a given methodology or collection of methods. Instead, this volume takes a rounded view of the various fields and perspectives that have become established in visual communication scholarship, offering both students and academics a comprehensive overview supported by empirical research.

In doing so, in the book we focus on the major contributions that media and communication scholars have made to the study of visual communication. We bring together important debates across media and communication studies in three key areas of enquiry – namely, **identities**, **politics** and **commodities**. We also use these three thematic strands to organize the book in three main sections. As we summarize below in relation to the book's structure, another key distinguishing feature of this volume is the inclusion of 18 original case studies, which are equally distributed across its three sections. Through these case studies, we offer a wide range of compact analyses that are both visually led and richly contextualized. Within each of the case studies we apply a variety of research approaches and methods appropriate for answering specific questions about how aspects of the visual are imbricated in mediated encounters with contemporary social issues.

SO HOW DO WE PROPOSE TO ENHANCE UNDERSTANDING OF CONTEMPORARY VISUAL COMMUNICATION?

To say a little more about why this is a fascinating time to be studying media images, we would point to three trending concerns for visual communication. First, the idea that we are 'bombarded' with images in our daily lives is much repeated, and while there is a sense that it has become easier to produce, share and look at images, along with other visual information, we caution against generalized assumptions about the nature of attention and the emotional investment afforded in such characterizations of image abundance. It is only by conducting empirical work – ranging from gathering evidence on how visual images are framed or shared to talking to people who make or use imagery – that we can evaluate the significance of visual communication practices in different contexts. For this reason, in this book we use a range of approaches and methods in order to capture a rich understanding of perspectives, motivations and tensions across varied forms of mediated images.

Second, there has been intense debate around the diminishing authority of the image, and especially its truth-value when it comes to documenting social reality; manipulative practices are as old as photography itself, but the ease and accessibility of digital technologies offer a profound challenge to the evidential value of photojournalism and documentary film. However, in paradoxical fashion, more of life's events have become worth recording and sharing, often via mobile devices, and with occasional (and even accidental) eyewitness journalism bringing attention to social and political injustices. This paradox means that we need to examine how images are valued, and through which means they acquire iconic status or become symbolic currency in a variety of communicative arenas, including but not limited to social networking, political communication and promotional culture. How are image-making activities and the resulting images spoken about, and what are the distinctive thematic or stylistic features of such images?

Third, images have become a fundamental part of how we communicate, self-identify and recognize our like-minded peers, especially via digital and mobile technologies. Online and especially 'social' media technologies have attracted a lion's share of scholarly attention in recent years, with a focus on self-representation in digital culture (Thumim, 2012). To neglect the meaning-making potential of visual imagery in this environment is to disregard vital and complex representational practices in our 'semiotic landscape' (Kress and van Leeuwen, 2006; Iedema, 2003).

HOW ARE WE DEFINING VISUAL COMMUNICATION AND WHAT DO WE MEAN BY 'UNDERSTANDING IMAGES IN MEDIA CULTURE'?

In its broadest sense, visual communication encompasses all forms of communication using the sense of sight. Theories of perception and vision investigate the physiology of the eye, and how psychological and social factors affect what we see. Shifting attention from the perspective of ocular perception to materiality, visual design, product packaging or physical space are also major aspects of visual communication. But our focus is more specifically on the images that circulate in the media, and by this we mean the non-linguistic pictorial elements that feature in cultural artefacts distributed via media technologies. Images are produced via various means (drawing, photography, animation, digital effects) and so their 'content' should be considered alongside their form and medium. The choices made regarding representational form and content work together to produce meanings through visual means (in interplay with other modes of communication). However, that is not to assume that meaning is completely fixed in the visual media object; rather, meaning potential is only realized in the specific social settings and through the discourses that contextualize the image.

Much has been written of the 'polysemic' nature of images or their capacity to have several possible meanings. This is often expressed in the distinction that images 'show' rather than 'tell', and that they are more 'open' as texts. But a plurality of possible meanings does not necessarily imply that the meanings of images are entirely free-floating or only determined in the eye of the beholder. Any effective form of communication relies on messages being understood in a particular context: sign-makers choose the most apt and plausible elements and forms in order to effectively represent the intended message (Kress and van Leeuwen, 2006). If images are so ineffectual in communicating ideas, knowledge and values, how do we explain their abundance in media culture? It is worthwhile, then, to examine how images convey certain intended meanings in the visual strategies employed by producers. But it is not only their intended meanings we are interested in; it is also the way in which visual images reveal latent values and underlying ideologies in their codes, aesthetics and rhetorics. However, to 'make sense' of the ideas, events and feelings expressed and exchanged through media images, a shared cultural understanding is needed at some level (Hall, 1997).

This said, as Raymond Williams (1976) famously claimed, culture is one of the two or three most complicated words in the English language, and different traditions of learning provide divergent understandings of the term. The concept is also often coupled with a prefix such as 'popular', 'national', 'counter' or 'political' to indicate 'the meanings and values surrounding a given activity or sphere' (Richardson et al., 2012: 4). As Richardson et al. (2012) further elaborate, the 'leakage' between the two broad meanings of culture (as associated with arts and expressive activity, or as delineating the boundaries of shared social values and meanings) can hold strategic convenience for those using the term. Indeed, there is a certain woolliness to 'media culture' that allows us to embrace a diffuse set of products, practices and technologies, while also encompassing a wide set of social values and aesthetic forms.

For our purposes, media culture refers to the sites, technologies and practices where meanings circulate within both traditional 'mass' media and personalized media forms (such as social networks). Media culture comprises the cultural artefacts that embody a series of representational choices made by producers, the manner in which they circulate via media technologies and the audience or user interactions when engaging with such media products. Finally, we should not forget that media culture also exists within certain economic and political structures, and that it is therefore shaped by patterns of ownership and platform dominance which undoubtedly produce imbalances when it comes to representational power.

In our quest to understand 'images' in 'media culture', we tend to privilege the site of the image as a trace or record of wider cultural values, activities and norms. So while the majority of our case studies focus attention on the formal characteristics and the patterns of representation of a selection of visual images, we also include interviews with producers and audience research (see Chapter 2 for further details). Even in narrowing down our perspective on visual communication to the pictorial elements of artefacts distributed via media technologies, the complex and multifarious nature of such media images requires us to be selective in what we are able to cover in this book. The next section briefly 'maps' the field of visual communication studies. We then clarify our own approach to visual communication before turning to the three main thematic strands we use to organize this book.

MAPPING VISUAL COMMUNICATION STUDIES

In comparison to textual communication, and positioned within the broader discipline of media and communication studies, we can argue that visual communication has been understudied. But from an alternative perspective, there are a number of academic fields of study where the visual has long been a central concern: art history, film studies and photography theory being the most prominent ones. Even in such image-centred fields, scholars have relied on words in their efforts to describe and translate the qualities or essence of the visual form. Although visual essay formats have started to appear in a limited number of academic journals, we usually rely on words to talk and write about visual arts and media images.

In addition, there are disciplinary traditions where the visual has perhaps been considered a peripheral 'wing' to the primary research questions and modes of enquiry of a number of fields and subfields (sociology, anthropology, political communication, journalism studies, international relations, cultural studies, computer science). In some of those cases, visually focused research can now be considered a subfield in its own right. For example, both visual sociology and visual anthropology offer methodological instruments to conduct ethnographic fieldwork with the visual aid of photography and video, but also drawings and maps, among other types of imagery (Stanczak, 2007; Pink, 2013; Pauwels, 2015a and 2015b). These are typically images that are made by the researcher or research participants to address research questions regarding the social relations, identities or structures that set apart a particular research site. Sometimes pre-existing or 'found images' (e.g. family photographs or archival images) are also used to elicit memories and responses during interviews with research participants. In both cases, visual images are a means to an end, rather than the main research focus. While there are significant overlaps between this type of visual research and research on visual communication, in this book we do not consider approaches and analyses where images are used as 'one type of data amongst others generated usually by interviews or ethnographic fieldwork' (Rose, 2016: 307). Instead, here we focus deliberately on methods and approaches that contribute to an understanding of images as our main objects of study, and therefore also on visual communication as a transdisciplinary yet cohesive field.

It is possible to chart the recognition of visual communication as a field in its own right by taking a close look at the interest groups and divisions of major international associations. For example, the Visual Communication Studies division of the International Communication Association (ICA) was originally founded in 1993 as an interest group that brought together those already integrating serious study of the visual into their work in anthropology, politics and journalism in particular (Barnhurst et al., 2004). The European Communication Research and Education Association (ECREA) only gained a Visual Cultures working group in 2015, although this is a much younger association, coming into existence itself in 2005. As Marion Müller (2007) argues, the work of German immigrant cultural historians such as Erwin Panofsky, in designing standardized methods of visual interpretation, provided an early bridge between art history and the social sciences in the first half of the twentieth century. In her article,

'What is visual communication?', Müller argues that the transdisciplinary nature of visual communication is both a positive and negative quality: the heterogeneity of relevant traditions offers scope for visual researchers, but it has also meant that visual communication has suffered in terms of recognition for funding and curricula. Müller's overview of visual communication as a research field follows an earlier mapping article by Kevin Barnhurst and co-authors (Barnhurst et al., 2004), and we recommend these detailed and vivid accounts of both Anglo-American and European contributions to the field.

Another marker of a maturing field is the development of academic journals. The Visual Communication division of the Association for Education in Journalism and Mass Communication (AEJMC) produce their own journal, *Visual Communication Quarterly* (since 1994), while the SAGE journal, *Visual Communication*, welcomed a new editorial team in 2018 – the first changeover since its launch in 2002. The new editors point out how 'changes in technology have made more manifest and tangible the inherent multimodality of all communication' (Ravelli et al., 2018: 398), thus also leading to important transformations in academic research agendas and theories of the visual. Since 2002, SAGE has also published the *Journal of Visual Culture*, with recent special issues covering diverse themes such as visual activism, internet memes and architecture. Possibly even more prone to existential angst than their visual communication cousins, visual culture scholars have provided critiques that question the very nature of the 'field', its objects of study, interpretive methods, boundaries and definitions (Mitchell, 2002; Moxey, 2008).

As with visual communication, the dropping off over the past 15 years of articles attempting to define the field could be read as a sign of growing intellectual confidence. We do not intend to mention all relevant journals here, but note a healthy scene of publication opportunities for critical analysis, creative practice and theory building across specific visual communication and culture journals, in addition to the more general media and communication or cultural studies publications.

It is, then, becoming harder to argue that there is a general neglect of the visual. There are still cases of book or article titles containing the words 'visual' or 'image' that do not necessarily examine the nature of the visual in any serious detail. But a scan of media and communication journals shows the attraction for articles about, say, the production of news photography, the sharing of social media images or the ways in which audiences respond to visual messages. Visual communication research appears to be much less 'niche' than it was at the beginning of this century. To some degree this follows the identified 'visual turn' across the humanities and social sciences – that is, an increasing willingness by scholars to pay attention to images and other visual forms, and to approach such images as socially constructed and as constituted dynamically through their relationships to other images, texts, viewers, users and contexts.

OUR APPROACH TO STUDYING VISUAL COMMUNICATION

What we have selected to cover in this book undoubtedly stems from our own personal scholarly histories and interests, as well as our current working location in Leeds in the United Kingdom. With a background in semiotics and both communication and cultural studies, Giorgia's past work has focused on the relationship between visual communication practices and the representation and promotion of particular identities. Her social semiotic research on the visual communication of European identity, for example, combined ethnographic fieldwork with image-makers, including graphic designers and photographers, with a detailed analysis of images' visual resources (Aiello, 2012a; Aiello, 2012b). Her more recent work on stock photography and generic imagery has integrated considerations regarding the digital presentation and circulation of images into a critical understanding of their meanings (Aiello and Woodhouse, 2016; Aiello, 2016). Coming from a politics and communication studies background, Katy's past work has drawn upon traditions such as news framing, while placing the visual as central to questions of representation and the construction of social realities. For example, her studies of the visual framing of the 2003 Iraq invasion and the 2006 Israel–Lebanon conflict combine quantitative content analysis of news photography with qualitative semiotic readings (Parry, 2010a; Parry, 2012). More recently, the visual expressions of political affinities on social media and the nature of political visibility for varied actors have become other relevant areas of interest (Parry, 2015; Parry, 2019).

It is also and foremost thanks to our combined strengths that, throughout the book, we are able to use varied approaches, research questions and methods in our original case studies. What the chapters do share is a social constructivist perspective – an understanding that images play a significant role in constituting the world around us and that this is a dynamic relationship. As David Machin and Andrea Mayr (2012: 10) succinctly express it: 'Visual communication, as well as language, both *shapes* and *is shaped by* society.' We are interested in the meaning-making potentials of semiotic resources, rhetorics and discourses, but in each case, whatever the particular tools or terminology of analysis, we situate our observations and interpretation within the broader contexts in which images are made, shared and used.

Our interest is not only in the features of visual design; it is also in the social, cultural and political significance of media images. Stuart Hall's writing has been a formative influence on us both. Likewise, the critical philosophies of figures such as Roland Barthes and John Berger have infused our own thinking on photography and the politics of representation. Moreover, Gunther Kress and Theo van Leeuwen's social semiotic framework in *Reading Images* (2006) has undoubtedly shaped both of our outlooks on visual analysis as a powerful means for an investigation of culture, politics and society through the lens of aesthetics and formal detail. We say more about visual communication as a field and visual methods in the next chapter, but we would also add that the multiple editions of Gillian Rose's *Visual Methodologies* (2016) have been invaluable for our research and teaching.

Before moving on to discuss key methodological approaches in the next chapter, we would like to briefly recognize a selection of major works that have inspired us within this field. As media and communication scholars, we would especially like to stress our indebtedness to the work of photography theorists and embrace the importance of bearing in mind 'the specificity of the medium, and the practices built around it in social use, where signification actually takes place' (Lister and Wells, 2001: 73). Indeed, we would apply this rule to any of the visual media formats and genres we discuss. Photography as a medium has attracted an especially vast array of scholarship which continues to battle with its ontological status and indeed its apparent resurrection after announcements of its death or the post-photographic era.

The theoretical work on photography and broader visual practices by writers such as Roland Barthes (1982), Walter Benjamin (1968), John Berger (1980) and Susan Sontag (1979) is embedded in our approach. The relationship of photographs to truth and authenticity continues to haunt how we view photographic images. The evidential and 'objective' qualities of the photograph are not only questioned due to its status as a trace of institutional and ideological discourse (Tagg, 1988), but also due to the growing ease with which images are now staged, faked, decontextualized, recontextualized and mislabelled (Ritchin, 2009).

The capacity of photography to induce empathy for distant others continues to fascinate in broader ethical debates about the representation of suffering (Linfield, 2010), the photograph's relationship to memory (Zelizer, 1998) and, for Robert Hariman and John Louis Lucaites (2016), also how the photograph can perform as a 'public image' in the sense of realizing collective ideals. At the same time, the development of photography into a veritable global cultural industry has perhaps led to a weakening of its representational and performative status. Both 'specific' editorial images and 'generic' stock images are now exchanged as commodities in an increasingly saturated market (Frosh, 2003). Here as well, the complexities of the new media environment and competition from social media images are testing what we understand and define as 'news' and 'editorial' or, on the other hand, 'commercial' and 'stock' photography – with photographers facing challenging conditions in a system where the economic and cultural value of their work is increasingly detached from the materiality of their main output – namely, the photographic image.

The contributions we have just covered, and photography theory as a whole, are not so easily discussed in terms of research methods (to which we turn in Chapter 2). More broadly, photography and visual culture scholars like Martin Lister and Liz Wells (2001: 73) have commented that the cultural studies analyst is wary of the term 'communication', preferring 'representation' or 'mediation' due to the former's association with more 'rigid application of systematic methodologies'. Overall, this book is centred in media and communication studies approaches that combine methodological insight with an appreciation for the importance of theoretical grounding and interpretive nuance.

We therefore position ourselves primarily as visual communication scholars following the traditions of media and communication studies, rather than a visual culture tradition which stems more properly from art history, film studies and cultural studies. This is, however, a matter of emphasis rather than clear boundaries: both traditions scrutinize the cultural and social implications of visual image-making practices; the differences are found in the chosen objects of study, preferred theoretical framings and modes of enquiry.

Something else that can be found throughout this book is a general framework explicitly set out in a three-pronged approach to visual analysis. As a whole, we structure each of our case studies in three interlinked analytical stages: *descriptive* analysis, which identifies the visual resources, codes or signs within and across the corpus of texts; *interpretive* analysis, which considers the different ways in which meaning potentials are established by situating visual resources in their specific historical, cultural or institutional contexts of production and use; and finally, *critical* analysis, which links texts and contexts in order to understand the cultural significance and ideological import of imagery together with the power-laden implications of visual communication for politics and society more broadly (Thurlow and Aiello, 2007). These stages are followed in a systematic manner by social semioticians (van Leeuwen, 2005), but we can also note resonances of earlier analytical distinctions – for example, in the categories of denotation, connotation, and myth applied in Roland Barthes' structural semiotic analysis (Barthes, [1961–73] 1977; Barthes, [1964] 1977). Critical discourse analysis also often works through a series of questions that move from representational forms to the identities and relationships 'set up' in the media discourse (Fairclough, 1995b), to the forms of knowledge and legitimacy being promoted, and the way in which power relations are reproduced in society.

It is worth saying a bit more about the 'critical' analysis stage here. For us, this is about the nature of the questions that underpin any research and moving beyond identifying certain features or characteristics to ask, in the words of David Machin and Andrea Mayr, '*why* and *how* these features are produced and what possible ideological goals they might serve' (2012: 5). While a desire to reveal power relations and expose their naturalization undergirds critical research, we would also warn against a critical approach that transmutes into one-track 'criticism'. We follow Gillian Rose's (2016: 22) call for a critical approach that 'takes images seriously' by looking carefully at the images themselves, thinking about the wider social conditions and cultural practices, and reflecting on our own ways of looking. Being critical is not about applying a predictable formula whereby power or cultural hegemony can be identified and admonished, but remaining curious, invested and questioning of the apparent incongruities and contradictions that in-depth analysis reveals.

WHAT WE COVER IN THIS BOOK (AND WHAT WE DON'T)

This book covers an eclectic array of media formats and genres, and yet there is still a range of media forms not covered. As outlined above, we both hail from certain academic traditions, and so the book will be inevitably flavoured with our own specific research interests. There is certainly a diversity of visual media images examined in this book. Within a broad category of photography, we include photojournalistic images in a news context, photographs shared on social media, stock photography in image banks, advertising images and promotional media. Differentiating moving images, we include YouTube videos, television title sequences, film trailers, and both charity and advertising campaigns. In addition to this, we analyse the visual properties of websites, internet memes, protest materials and hashtag campaigns. We provide further details in Chapter 2 on the methods we apply in each case study, but we selectively employ techniques from semiotics, multimodal discourse analysis, visual rhetoric, visual framing and content analysis, in addition to conducting interviews, surveys and focus groups.

Even with the diversity of media formats and methodological approaches outlined above, it is necessary to be selective. In some cases, this is due to the particular media genres and media formats having their own distinctive analytical tradition which falls outside our immediate focus. For example, we have included film trailers but not covered film as a medium in any detail. This is partly due to the existing abundance of film analysis guides, together with the fact that film studies has developed as its own discipline well beyond the boundaries of media and communication studies as such. We also do not deal with video games, infographics or artworks, as they fall outside our own primary areas of interest and expertise. The simple truth is that we cannot cover everything in a single volume, but we hope to point the reader in the direction of relevant and insightful studies of visual communication.

In the next section, we turn to an explanation of the issues at the heart of each of these three main sections together with a chapter overview.

STRUCTURE OF THE BOOK AND CHAPTER OVERVIEW

We consider that the structure of this book is quite unique. The book is divided into three main sections – **identities**, **politics**, **commodities** – representing the three major and overlapping strands of enquiry outlined earlier. Each section includes three chapters, providing a theoretical and empirical toolkit for visual communication research and visual analysis. A major distinguishing feature of this book is that it covers a discussion of 18 original case studies, which are equally distributed across its three main sections. The chapters each provide an introductory overview of the relevant scholarly debates and then include two case studies. In addition to contributing to our three main areas of enquiry, the case studies in each chapter are designed to function as blueprints for research on related topics.

We have carefully chosen case studies that illuminate significant features of visual communication across a range of media genres and forms. In some cases, these are selected due to their emblematic status or apparent iconic power, but others represent the everydayness of imaging practices. The subjects of our 'snapshot' analyses are designed to serve as a springboard for reflection on the larger issues of visual images in today's media culture.

To guide our hands-on analyses of case studies, in the next chapter (Chapter 2) we introduce a range of methods for visual analysis, with the fundamental understanding that our choices of methods ought to be dictated by the questions that we have about the relationship between a specific set or type of images, and a meaningful aspect of cultural, social and/or political life.

The book's first main section (Chapters 3, 4, 5) focuses on how **identities** are shaped and maintained through visuals aimed at performing the self, communicating a collective identity or foregrounding the differences of others. This section of the book examines both burgeoning and long-standing aspects of media culture like self-representation, national image-building and stereotyping.

Starting with practices of self-representation, Chapter 3 examines the ways in which images are produced, circulated and used by 'ordinary people', and how this relates to identity formation and negotiation. With the introduction of widely available image-making devices and image-sharing platforms, individuals have gained unprecedented access to forms of self-representation previously mediated by powerful institutions like television networks and museums. Social networking sites like Facebook, Instagram and YouTube have contributed to the development of visual subjectivities that actively combine expressive intents, interpersonal uses and promotional goals. By focusing on the relationship between the visual and the self, this chapter explores the tensions and contradictions that are typical of digital self-portraiture and online micro-celebrity. The chapter's case studies cover selfies on Instagram and YouTube celebrity video tutorials. In both cases, we seek to examine the visual resources employed by image-makers and viewers alike to 'make sense' of this imagery as authentic, credible and appealing.

Chapter 4 focuses on the promotion of collective identities in the mediatized arenas of television, public relations and social media. We begin by discussing the significance of visual communication for national image-building practices that range from participation in mega-events to nation branding. We also consider the increasing visibility of collective identities that move beyond the nation. The chapter balances a critical stance on the visual communication of specific and potentially exclusive versions of shared identity with an appreciation of the strategies used to craft novel visions of community. The case studies focus on the storytelling and rhetorical devices used by the Everyday Africa Instagram feed and promotional media produced by cities participating in the European Capital of Culture scheme, respectively.

The final chapter in this section focuses on the defining power of the media to constitute the 'other' through an emphasis on difference and familiar stereotypes. Chapter 5 considers the writing on Orientalism, othering and stereotypes with an emphasis on how the *pictorial* aspects of representation hold particular power to injure, even if not deliberately intended. We also consider attempts to move past such simplifying tendencies – for example, in the desire to harness the humanizing potential of photography and film to connect with distant others and appeal to a universal moral cause. We touch on issues of ethnicity, class, gender and inequality within this chapter, and present two case studies where visual strategies are employed with the clear intent to move beyond 'othering' practices. We analyse the television title sequence of the hit Netflix show *Orange is the New Black* and a charity appeal on behalf of the Rohingya refugees by Save the Children.

The book's second section (Chapters 6, 7, 8) focuses on **politics**, an area that has become paramount to an investigation of visual images and which is here broadly conceived to include mainstream, activist and international relations outlooks on visual communication in the public sphere. This section considers how politicians, and especially female politicians, face challenges in managing their visibility across the media ecology, the circulation of activist imagery of protest and the news mediation of international conflict.

Chapter 6 begins with the world of 'official' politics, explaining why traditionally there has been an uneasy relationship between the political realm and the visual. We examine how politicians have sought to control their image and visibility in varied media forms, and how digital technologies offer both potential opportunities and pitfalls. We explore the gendering of politics and how female politicians especially face challenges in both gaining media visibility and political legitimacy. The case studies include an analysis of the 'visual cues' found in the gendered representation of female politicians, with a focus on front-page newspaper covers of Theresa May during Brexit negotiations, and a look into the dark side of meme culture and virality with an analysis of the visual rhetoric of disinformation found in the Russia-linked Facebook adverts from the 2016 US presidential election.

Chapter 7 shifts focus from 'top-down' to 'bottom-up' politics, considering how protesters harness the rhetorical power of the visual to fight the powerful. In order to gain visibility, activists and social movements often exploit the carnivalesque and theatrical to attract media attention, using various forms of 'media', from graffiti to YouTube videos. This chapter explores three levels of visual communication in this mix: the visual materials used by activists during events; the mass media representation of protest and dissent; and the 'self-mediation' used by activists to raise awareness and mobilize support. The case studies apply semiotic analysis to explore the visual techniques deployed by Russian punk collective Pussy Riot and examine how a Black Lives Matter 'visual icon' is co-created via Twitter images.

Chapter 8 explores how images of war hold the potential to convey some of the most devastating consequences for humans living through violent conflicts, but also how such images are filtered through various mediating processes, whether journalistic practices and conventions, or via the personalized spaces of social media platforms. This chapter summarizes the role of the image in portraying war and explains why images of conflict offer such potent 'visual icons'. The case studies include a visual framing analysis of ISIS propaganda in the UK press and analysis of survey responses to an iconic image from the Syrian war: Omran Daqneesh, the boy in the ambulance.

The third section (Chapters 9, 10, 11) concerns the promotion of material and immaterial **commodities**, including commercial products, consumer lifestyles and corporate brands. From an analysis of traditional forms of advertising and commercial imagery, to an investigation of how images contribute to experiential forms of consumption, this section offers an up-to-date take on a classic area of visual communication scholarship.

Chapter 9 considers the history of images and commodity culture. The chapter explains how promotional culture, advertising and capitalism sustain each other, and explores the history of critical analysis of advertising and promotion. Specific to visual communication, we show how semiotic analysis helps to reveal the codes and conventions of advertising and how certain ideologies become naturalized in such codes. For the case studies, we conduct an analysis of the O2 'Be more dog' advertising campaign to explore the appeal and humour of animals, and reveal how film trailers combine rhetorical appeals of genre, story and stardom through an analysis of the *Black Panther* (2018) trailer.

In Chapter 10 we address how in contemporary media it is not only products that are sold but also lifestyles, or ways of being that are intimately tied to consumer culture. The chapter considers the increasing commodification of everyday lifestyles more generally and lifestyle politics more specifically across different media outlets and media genres. We demonstrate the central role that visual images play in the promotion of particular ways of life pertaining to choices in matters of health or leisure, but also political outlooks. This chapter's case studies are a multimodal analysis and political economy critique of Vice Media's countercultural lifestyle aesthetic, and an investigation of 'feminist' stock imagery in the Lean In Collection by Getty Images, which combines considerations about images' production, representation, and both circulation and uses.

Our final chapter in this section considers brands as visual experiences. Chapter 11 looks at branding as a series of techniques used to strategically communicate the uniqueness and appeal of both material and immaterial commodities. The chapter considers the significance of promotional culture, and brands in particular, in how we experience and relate to various aspects of contemporary social life. Here we foreground the role of imagery and visual aesthetics more generally in creating distinctive brand identities and lasting bonds between consumers and a range of products, services and institutions. For the case studies, we examine changes in the visual branding of Airbnb over time, and explore the reception of logos of popular digital media technology brands like Instagram, Apple and Facebook.

We end with a short conclusion (Chapter 12) which reflects on the core aim of the book – that is, to offer an up-to-date, research-led guide to visual analysis in media and communication. We highlight the need to keep researching visual dimensions across identities, politics and commodities, together with the many intersections between these strands of enquiry.

Taken together, the book's chapters offer a combination of timely topics, methodological diversity and analytical breadth. As a whole, this is not so much a 'how to' methods book, but instead follows a hands-on research approach that emphasizes how images are put to use within certain contexts and through particular media. Overall, the book not only offers an account of the textual characteristics of visual communication, but takes into consideration the technologies and industries that are behind images, and the views of producers and aspects of reception, thus successfully engaging with the various 'sites' that shape the meanings and overall lives of visual images in media culture (Rose, 2016).

2 UNDERSTANDING IMAGES IN MEDIA CULTURE: METHODOLOGICAL CONSIDERATIONS

This book is not primarily a textbook focusing on methods for visual communication research, but in our approach we include original case studies to illustrate various applications of methodological approaches and instruments for understanding the 'work' of media images in specific social contexts. In the 18 case studies presented in this book, we aim to cut through the sometimes obfuscatory terminology to demonstrate how a chosen methodological approach offers a particular toolkit to scrutinize the role of images within each media 'recipe'. Rather than a one-size-fits-all approach, we mix and match methods and approaches to address specific research questions and to highlight both their strengths and limitations. For these reasons, in this chapter we will:

- outline a broad set of principles to guide the selection of methodological instruments for visual communication research;
- summarize and discuss key contributions originating from six major methodological approaches to visual analysis: content analysis; semiotics; critical discourse analysis; social semiotics; visual rhetoric; and visual framing;
- conclude with a brief discussion of some of the methods that are typically used to investigate the production, reception and circulation of visual images, including interviews with image-makers, visual elicitation techniques and digital methods.

We would not be the first to argue that visual analysis benefits most from a combination of complementary methods and that the choice of a particular methodological approach is intimately tied to the research questions that drive the study at hand, to the extent that scholars trained in different methodological traditions will often ask substantially different questions about visual images (van Leeuwen and Jewitt, 2001; Rose, 2016).

However, our approach here is distinctive because, rather than focusing on methods for visual analysis as such, our book provides compact visual analyses that illustrate multiple vantage points for answering critical questions about media images. Through its selection of case studies, the book introduces ideas, options and strategies for question-driven research in key areas of visual communication scholarship.

HOW TO CHOOSE METHODS FOR VISUAL ANALYSIS: SOME KEY TENETS

In our teaching we find that students often get confused about methods, especially where the terminology is obscure or complex. Even determining what actually counts as a method as opposed to a general methodological approach or framework can be challenging. For example, critical discourse analysis is not a single method that you can 'apply' or 'use' to collect or analyse data; rather, there is a diverse range of methods that could be adopted in studies following a critical discourse analysis approach. As Ruth Wodak and Michael Meyer (2016: 3) explain, to do critical discourse analysis and, we argue, also visual analysis more broadly, you need to begin 'by formulating critical goals, and then explain by what specific explicit methods you want to realize it'. Indeed, Teun van Dijk (2013) recommends the term 'critical discourse studies' for the 'theories, methods, analyses, application and other practices of critical discourse analysts'. Therefore, we would second van Dijk's advice that:

> A good method is a method that is able to give a satisfactory (reliable, relevant, etc.) answer to the questions of a research project. It depends on one's aims, expertise, time and goal, and the kind of data that can or must be generated – that is, on the context of a research project. (van Dijk, 2013)

While van Dijk is referring to critical discourse studies specifically here, this advice stands for most research projects. Thinking about research goals and how best to achieve them by selecting the most useful tools and analytical techniques can hopefully help both students and researchers to avoid anxieties about varied methodological approaches that come with their own confusing terminology and yet often appear to address a similar set of questions, at least when focused on media texts. For this reason, we also like the three questions about 'media output' posed by Norman Fairclough in his classic text *Media Discourse* (1995b: 5) as a starting point for analysis:

1. How is the world (events, relationships, etc.) represented?
2. What identities are set up for those involved in the programme or story?
3. What relationships are set up between those involved?

In the shortened form, Fairclough uses the categories *representations*, *identities* and *relations* to guide interrogation about media texts, and argues that these would be applicable to any media genre. The three questions relate specifically to 'media output' and so are directed at what Gillian Rose (2016) calls the 'site of the image'. For Rose, however, this is just one of the sites 'at which the meanings of an image are made' (2016: 24). Rose's framework draws upon established models of communication in that it identifies the site of production, the site of the image, the site of circulation and the site of audiencing as spheres where 'the meanings of an image are made' (Rose, 2016: 24) and where the social and cultural impact of visual communication can be studied.

The four sites (production, image, circulation, audiencing) are certainly most useful when it comes to differentiating a primary research focus and how a chosen method is therefore underpinned by assumptions about the importance of each site in the meaning-making process. In other words, as visual researchers we would not choose to focus our attention on the site of the image unless we thought that analysis of imagery's form and content might allow us to better understand something of the social implications of a particular type of visual communication. While, as we will explain in the next section, all major methodological approaches to visual analysis are indeed geared towards an in-depth investigation of the site of the image, some of the most nuanced analyses often combine methods to examine the form and content of images in combination with their production, circulation or audiencing (or reception).

As a whole, a theoretical understanding of the role of images in society is what drives the specific mode of enquiry. In addition to identifying the site (or sites) of study, Rose suggests that each site can also be critically examined by asking questions based on three aspects or 'modalities'.

The first is the *technological* modality, which focuses on how different visual technologies impact on its meanings and effects (what technologies are used to make and circulate the image); the second is the *compositional* modality, which refers to the material qualities of an image (colour, content, genre); and the third is the *social* modality, used by Rose (2016: 26) to refer to a 'range of economic, social and political relations, institutions and practices that surround an image and through which it is seen and used'. Rose offers her framework as a heuristic that enables the reader to locate the site/modality that they are most interested in, to then identify an appropriate method.

There is a second framework that we would like to acknowledge. In this case, we refer to the cultural studies tradition and the 'circuit of culture' first introduced by Paul du Gay, Stuart Hall and co-authors in *Doing Cultural Studies: The Story of the Sony Walkman* (1997). The 'circuit of culture' model has been used by many scholars and has been reproduced across key cultural studies texts. It proposes five elements for analysis of cultural artefacts: representation, production, identity, regulation and consumption. In telling the 'story' of the Sony Walkman – a portable cassette player with earphones – it was possible to trace how the Walkman was represented as a cultural artefact (in advertising, for example), how it was 'encoded' with meanings during the production process, how types of people became identified with its use, how it was shaped by the conditions of industry regulation, and the ways it was actively used and enjoyed by consumers (du Gay et al., 1997).

The Sony Walkman might appear to have little relevance to today's media landscape, but the 'circuit of culture' approach offers another framework to study the dynamic relationship between the different elements and processes in media culture. The critical goal driving this approach is how the interrogation of each of these dimensions enables the researcher to identify assumptions and hegemonic reproductions of a dominant culture and, conversely, the processes by which such assumptions or ideologies are contested. With his earlier article on 'encoding' and 'decoding' in television discourse, Hall ([1973] 1980) provided part of the inspiration for the model, alongside Raymond Williams's 'social' definition of culture. In turn, Hall's approach was inspired by Barthes' semiotic readings of photographs and adverts. Rather than studying individual acts of communication, Hall ([1973] 1980) applied a semiotic paradigm to mass media, conferring the 'decoder' with a significantly active role, leading to three possible types of reading: dominant or preferred, negotiated or oppositional. A quick scan of Google Scholar reveals the varied applications of the circuit of culture model in more recent years – a study of the Kinder egg, of public relations strategies in Brazil and of the promotion of Angelina Jolie's humanitarian activities, to name just a few.

Models such as those discussed above provide useful guidance for thinking analytically about the various elements in the meaning-making processes of media culture. The 'sites' or 'moments' of meaning are useful to differentiate for analytical purposes but, in reality, these are neither linear nor easily disentangled from each other. Nevertheless, such frameworks help us start to extricate how meaning is constituted and interpreted in different circumstances, and how potential meanings are organized through signifying practices. In the next section we look in more detail at methodological approaches designed to help us in this task.

METHODOLOGICAL APPROACHES TO VISUAL COMMUNICATION RESEARCH

To discuss the contrasts, similarities, strengths and weaknesses of the various methodologies, this section will summarize the various approaches and provide examples of key texts. Space dictates that those authors whose work deals specifically with visual communication research methods are favoured over others drawn from art history, film analysis or visual culture studies. For this reason, for example, influential works from the field of art history such as Panofsky's *Meaning in the Visual Arts* ([1955] 1987) and Gombrich's *Art and Illusion* (1960) are not detailed here.

Over the years, the production of textbooks on 'visual communication' has intensified, starting with the early editions of Paul Martin Lester's *Visual Communication: Images with Messages* (1995), A.A. Berger's *Seeing is Believing: An Introduction to Visual Communication* (1998), to then also include edited collections such as Theo van Leeuwen and Carey Jewitt's *Handbook of Visual Analysis* (2001), and more recent guides such as *Visual Communication Theory and Research: A Mass Communication Perspective* (Fahmy et al., 2014) and *Doing Visual Analysis: From Theory to Practice* (Ledin and Machin, 2018). The *SAGE Handbook of Visual Research Methods* (Margolis and Pauwels, 2011) is perhaps the weightiest in this selection, with over 700 pages devoted to a wide range of methods, including approaches to analysing visual images as well as visual methods for social research (e.g. photo diaries, rephotography and anthropological filmmaking). A new edition of this volume is due as we write, with a chapter on visual content analysis and another one on visual semiotics authored by Katy and Giorgia, respectively (Aiello, 2020; Parry, 2020).

We have already noted the importance of Gillian Rose's *Visual Methodologies* (2016) as a guide to methods as diverse as psychoanalysis, digital methods and photo-elicitation. Additionally, Stuart Hall's editions of *Representation: Cultural Representations and Signifying Practices*, with contributors updated in each edition, and Marita Sturken and Lisa Cartwright's *Practices of Looking: An Introduction to Visual Culture* (2017), now in its third edition, have provided crucial insights into the politics of representation and the ways in which mediated images offer artefacts through which we can analyse shifting ideologies of beauty, othering and identity construction. Gunther Kress and Theo van Leeuwen's *Reading Images: The Grammar of Visual Design* (2006) provides a classic guide to carrying out detailed and systematic analysis of visual design, and still stands as the blueprint for a social semiotic approach to visual and multimodal communication. We now present further details for the methodological approaches employed in this book.

Content analysis

As Anders Hansen and David Machin (2019) observe in their chapter on content analysis, this is one of the most efficient and widely used approaches in media and communication research. So, is content analysis simply anything that involves examining media content? Not really. A well-designed content analysis needs to follow a systematic and thorough approach, and is generally considered more 'scientific' than more interpretive methods of analysis, such as semiotics or discourse analysis. If you want to be able to make supported claims about the quantity or relative prominence of certain visual features, or to look for patterns across time or a diversity of media, then content analysis is a very useful technique.

Perhaps one of the most oft-cited definitions is Bernard Berelson's: 'Content analysis is a research technique for the objective, systematic, and quantitative description of the manifest content of communication' ([1952] 1971: 18). While some might take issue with whether any technique is entirely 'objective', the important takeaway words here are 'systematic' and 'manifest content'. Content analysis has to follow a series of steps in its design to ensure that it is measuring relevant content in a consistent manner, so a systematic approach is essential. Berelson's 'quantitative description' means that the content is summarized numerically (usually presented in tables of results): it is reported in aggregated form, rather than as a detailed discursive description.

A more recent and comprehensive definition of quantitative content analysis from Riffe et al.'s *Analyzing Media Messages* (2014: 19) moves beyond the scientific processes of this approach to take account of why we might want to conduct a content analysis: 'to describe the communication, draw inferences about its meaning, or infer from the communication to its context, both of production and consumption'. It is not simply about being able to describe the manifest content in summarized form based on a rigorous scientific process, but crucially it is about making inferences about communicative significance and interconnections with other sites of meaning (context, production and consumption).

There are a number of key methodological chapters for guidance when it comes to visual materials (Bell, 2001; Rose, 2016; Parry, 2020), but despite the presence of visual content in many forms of media, images remain on the periphery when it comes to quantitative methods handbooks. In keeping with its scientific traditions, there is a standard set of stages required in devising a content analysis and the chapters cited here provide guidance on how to ensure a robust research design. The positivist slant of content analysis and its emphasis on objective 'manifest' content raises concerns when visual images are the primary focus of study, as this is something that can feel restrictive when dealing with the ambiguity or communicative openness of visual images (Berelson, [1952] 1971). Here, it is worth thinking about how content analysis can work alongside other approaches to media analysis to make those 'inferences' about context suggested by Riffe et al. (2014) above. Supplemented with a rhetorical, discourse or framing analysis that aims to explore the thematic, strategic and persuasive elements of a set of visual images, content analysis provides the building blocks for a more sophisticated approach.

Two of the most influential visual content analysis studies are Erving Goffman's *Gender Advertisements* (1976) and Catherine A. Lutz and Jane L. Collins's *Reading National Geographic* (1993). While Goffman's characteristics of gender stereotyping, such as 'the feminine touch', have remained prominent in later studies on advertising, he has come under criticism for the rather opaque selection process of the chosen adverts (see Bell and Milic, 2002). But Lutz and Collins (1993) present a relatively clear rationale, selection process and coding strategy for their study of *National Geographic* magazine, covering from 1950 to 1986. With an interest in how 'cultural difference' between Westerners and non-Westerners was constructed in photographs from around the world, the content analysis revealed which regions of the world were favoured in reporting and how this shifted over time. Using variables such as 'world location', 'camera gaze of person photographed', 'urban versus rural setting', 'male/female nudity' and 'vantage', Lutz and Collins (1993: 285) were able to build a picture of the patterns of representations, and especially how troubling concepts of race and gender were perpetuated. The authors' analysis is also enriched by the inclusion of institutional and audience perspectives gained through interviews with magazine editors and readers.

In sum, content analysis is an excellent tool for telling us 'what is there' in a large corpus of media content (see an example of a simple content analysis in

Chapter 10's case study on feminist stock photos). But in order to both devise meaningful categories in the research design and enable interpretation of why such resulting frequencies or omissions are culturally or politically significant, the researcher needs to draw upon in-depth contextual understanding and relevant theories regarding the politics of representation.

Semiotics

Semiotics is concerned with how meaning is made and, in its most literal sense, it is the study of anything that can be taken as a sign (the Greek word *sēmeion* means 'sign'). Anything can be a sign as long as someone or, more importantly, a group of people who are part of the same culture or society interprets it as 'signifying' something – that is, as referring to or *standing for* something other than itself (see Bal and Bryson, 1991).

Building on Ferdinand de Saussure's linguistic theory of signs (de Saussure, [1916] 1983), in the 1960s Roland Barthes was the first semiotician, or *semiologist*, to focus on visual images, though he also applied a semiological lens to the study of fashion, cities, music and a range of popular 'icons' of French culture, including, among others, the Citroën car and the Eiffel Tower (Barthes, 1972, 1979). Whereas Saussure had looked at meaning-making in a *synchronic* manner (as if frozen in time), Barthes was interested in how meanings change across cultural and historical contexts, with a particular focus on how ideology works through images or other semiotic artefacts which are typically taken for granted.

With his theory on the layering of visual meaning, Barthes laid the foundations for semiotic and cultural studies approaches to visual communication as we know them today. In 'Rhetoric of the Image' (Barthes, [1961–73] 1977), he claimed that visual meaning can be articulated into the two separate levels of *denotation* and *connotation*. A now classic example used by Barthes in the same essay is that of an ad for a pasta brand. The denotative meaning (which is difficult to simply describe without adding connotation) of the image used in the Pasta Panzani ad is the immediate or 'literal' meaning relating to what is represented in the image – namely, a fishnet shopping bag full of packaged pasta, canned tomato sauce, onions, peppers and mushrooms together with a package of grated cheese, a tomato and a mushroom next to the bag – all of this being displayed against a red background.

The connotative meaning of this ad, instead, is one of *Italianicity*. Barthes points out that this ideological association between a simple shopping bag bursting with Mediterranean vegetables and pasta (along with the name Panzani, which is part of the ad's linguistic message) and the 'essence' of being Italian generally works for the French, whereas Italians might not even associate a connotation of *Italianicity* to this message. Connotation is therefore context-dependent and corresponds to the symbolic or ideological meaning, or range of possible meanings, of an image inscribed by cultural codes. Codes can be defined as the 'implicit rules' (Sturken and Cartwright, 2017) that govern the ways in which those who make and use images 'read' their meanings. The idea here is that, as part of a shared system of culture, most of us are able to draw from the same codes to interpret and understand images. Barthes' theory of meaning became central to the 'semiotic turn' in cultural studies, where his notion of connotation was taken as a springboard for an approach to visual analysis that emphasized the fact that images are polysemic as they 'do not possess a fixed or essential meaning' (Hall, 1997: 31). Therefore, meanings can be *unfixed*, since they are historically and culturally constructed, but ideology is also 'the power to signify events in a particular way' (Hall, 1982: 69).

While Barthes is without any doubt the most important founding figure in visual semiotics, another French semiotician, Jean-Marie Floch, was also central to its development. Floch's semiotic approach to visual communication was based on the distinction between *figurative* meaning and *plastic* meaning. Figurative meaning pertains to the representation of things and human beings, particularly in relation to the visual traits that enable us to recognize particular objects or subjects and the narratives that are associated with their representation (Floch, 1985). Floch's *figurative semiotics* is therefore germane to Barthes' approach to the 'rhetoric of the image'. Plastic meaning, on the other hand, relates to visual cues like line, shape, light, colour, texture and layout. These are all aspects of an image that can have meanings that are independent of what they represent from a figurative standpoint (Greimas et al., 1989). In his best-known book, *Visual Identities*, Floch ([1995] 2000) applies this approach to a series of groundbreaking studies of brands like Chanel, and both IBM and Apple. Overall, Floch developed a visual semiotics centred on the meaning-making properties of form – or both 'style' and 'design' (Aiello, 2007) – and its relationship with content.

As a whole, the main limitation of semiotics is that it is centred on detailed analysis of texts and their components, with little regard for the practices and processes that underlie their production or reception. Semiotic analyses are also often discounted as being merely individual readings of a particular text or set of texts rather than evidence-driven studies. It is, however, important to point out that Barthes' denotation/connotation model has contributed greatly to the development of critical approaches to the visual across a variety of disciplines. By the same token, semiotic analysis is also a great complement to more empirically grounded approaches and methods to examining meaning, including content analysis and interviews with producers or users of imagery, for example. Ultimately, a semiotic approach to visual analysis enables us to examine imagery both systematically and critically, therefore also combining 'analytical precision' (Rose, 2016: 107) with an ability to link textual details to both meaning and ideology (see Chapters 3 and 9 for semiotic analyses of selfies and advertising, respectively, and Chapter 11 for a reception-led semiotic analysis of the logos of a selection of global digital media and technology brands).

Critical discourse analysis

Critical discourse analysis stresses the historical moment of any interpretation and the constitutive nature of meaning through rules, practices and institutional power. As Weintraub (2009: 203) states, 'discourse analysis is a way to investigate issues of truth, power, and the social construction of reality'. Michel Foucault (1977) argued that discourse constructs the subject, but also that the way we share understanding and meanings is not a benign system of negotiation, but rather an apparatus in which power is inscribed in discursive practices. What we think we 'know' at a particular time in history, whether it is related to medicine or criminology – to use examples that interested Foucault – has implications in how such issues are defined and regulated.

Foucault's ideas on the role of discourse in linking power and knowledge have become incredibly influential in the social sciences, with varying strands of discourse analysis applied to verbal and visual texts as well as practices of institutions. For example, the importance of recognizing the particular context of news website images is reiterated in this approach – not only the historical specifics, but also the individual brand and the presumptions associated with it. Like any brand, each news website has a recognizable typeface and layout, all easily decoded visual elements that influence the expectations of the reader. A photograph displayed on one news website may hold a very different meaning when displayed in another, as it appears within a 'collection of various messages which surround the photograph' (Bignell, 2002: 95). In addition to the 'various messages' within the webpages of a single news site, knowledge of the broader context of the political and social environment, and the practices of production and dissemination of news, are also of significant benefit to the analyst.

Norman Fairclough developed a three-dimensional framework for studying the media in *Critical Discourse Analysis* (1995a) and *Media Discourse* (1995b), comprising *cultural practice*, *discourse practice* and *text*. In this regard, he stated that, 'on the one hand, processes of text production and interpretation are shaped by (and help shape) the nature of social practice, and on the other hand the production process shapes (and leaves "traces" in) the text, and the interpretative process operates upon "cues" in the text' (Fairclough, 1995a: 133).

In *Media Discourse*, however, the central focus is on the media text and the language used in media output. One strength of Fairclough's work is the rejection of a formalist approach to media language, a move away from isolated textual analysis (such as in traditional semiotics) to a more open, socially constitutive and intertextual conception of language and meaning. Rather than looking for what is 'there' in a supposedly objective textual analysis, intertextual analysis looks for the 'traces' of discourse practice in the text (Fairclough, 1995b: 61). For Fairclough, the 'text' includes 'visual images and sound effects' (for television) while 'discourse' includes both the term for 'social action and interaction', as defined in language studies, and the post-structuralist sense of discourse 'as a social construction of reality, a form of knowledge' (Fairclough, 1995b: 17–18). However, Fairclough's main focus is still on the linguistic element of texts rather than the visual.

Because critical discourse studies share a common interest in deconstructing ideologies and power through systematic investigation of semiotic data (Wodak and Meyer, 2016), the techniques of analysis associated with this approach are often employed in studies designed to interrogate representational features such as difference and othering in the media. According to this framework, then, we may want to ask the following questions:

- What particular visual features are used (recurrently) to identify a group?
- Are certain groups of people or individuals photographed in a positive or negative light? What activity are they shown to be involved in? Are they shown to have agency or as passive?
- Are they portrayed as symbols for a wider cause, morality or ideals?
- Do the choices in composition and framing imagine an empathetic or distanced spectator?

Thinking in terms of ideological discourse structures is a useful way to unpick how ideologies and attitudes become naturalized in visual communication, in addition to language use (see Chapter 5 for two case studies using a critical discourse approach to examine the representation of difference in the title sequence of the Netflix drama *Orange is the New Black* and in visual campaign strategies for the humanitarian organization Save the Children).

Social semiotics

Social semiotics extends some of the key instruments for visual analysis developed by Barthes, while also introducing important considerations about the significance of both context and practice for a semiotic understanding of the visual. Social semiotics originates from a synthesis of structuralism and Halliday's (1978, 1985) systemic functional linguistics. This approach is functionalist in that it foregrounds choice and considers all sign-making as having been developed to perform specific actions or *semiotic work*. Just like traditional semiotics, social semiotics is concerned with the internal structures of images and multimodal texts. Unlike traditional semiotics, and in line with critical discourse analysis, social semiotics places emphasis on the relationship between form and how people make signs 'in specific historical, cultural and institutional contexts, and how people talk about them in these contexts – plan them, teach them, justify them, critique them, etc.' (van Leeuwen, 2005: 3).

In *Reading Images*, Kress and van Leeuwen (2006) developed a highly influential social semiotic framework to analyse images according to three main *metafunctions* – namely, representational, interactive and compositional meaning. A social semiotic analysis of visual imagery will therefore typically address three main questions, whether implicitly or explicitly:

1. What is the *representational meaning* of an image or set of images? In other words, what is the 'story' that is represented? Who are the key 'participants' (the people or objects portrayed), how are they represented as types or as parts of a broader 'whole' and what are some of the actions that are performed by participants – for example, in relation to themselves or others, a certain product or their environment?
2. What is the *interactive meaning* of an image or set of images? In other words, how do the images interact with the viewer – for example, by means of a portrayed person's gaze, a certain camera angle, frame size, etc.?
3. What is the *compositional meaning* of an image or set of images? In other words, how are particular images laid out or organized – for example, through the ways in which different visual cues or objects are placed in the image, are made more or less salient, connected or disconnected, made to look more or less real, etc.?

Kress and van Leeuwen's framework is not unlike Barthes' denotation/connotation model, but the social semiotic approach considers a number of 'techniques' through which each of these levels of meaning may be established (for a more detailed explanation of the three metafunctions and the visual features that realize them together with their meaning potentials, see Kress and van Leeuwen, 2006). However, social semioticians have increasingly highlighted that using the three metafunctions is not a sufficient method to address the specificity and situatedness of visual images.

For this reason, a social semiotic analysis may also address four further aspects pertaining to the materials, uses, styles and practices that shape images' visual resources and meaning potentials (see Aiello, 2020 for a social semiotic analysis of a stock photograph based on this framework). First, one may want to consider the affordances that set apart the visual images being examined or the types of communicative acts and meanings that are enabled by their particular form of communication. For example, photography and film have material qualities and features (such as stillness vs. movement) that shape and make them apt for certain kinds of communication and interaction but not others (Ledin and Machin, 2018). Second, it may be useful to examine the canons of use associated with the images at hand by examining both the histories that shape the ways in which these images are used in specific contexts and the values that particular types of images are usually made to communicate (Ledin and Machin, 2018). Third, one may need to take into account the role that non-figurative elements or design resources like shape, light, colour, texture and layout play in shaping the style and overall content of imagery (see Floch, [1995] 2000) – something that is especially important for non-photographic and minimalistic images like logos (Johannessen, 2017). Finally, a social semiotic analysis should address the specific creative, professional or viewing practices that contribute to the visual resources and meaning potentials of particular visual texts – for example, through a review of existing literature, documents or trade media articles and/or, where possible, interviews and/or ethnographic fieldwork with key producers or users of imagery (see Aiello, 2012a).

As a whole, these are all analytical dimensions that can be addressed separately or, on the other hand, also go hand in hand. Depending on one's research questions, these different dimensions contribute to a social semiotic analysis of visual texts, as their materials, uses, styles and practices have all developed to accomplish communicative aims in the service of particular interests (see Chapters 7, 10 and 11 for social semiotic analyses of images of Pussy Riot, the rebranding of Airbnb, and Vice Media's official website, respectively).

Visual rhetoric

Visual rhetoric studies the relationship between visual images and persuasion. Visual rhetorical analysis typically identifies first the general type of rhetoric (such as a specific genre) and, within that, the rhetorical devices or techniques deployed (metaphors, symbols) and suggests how such qualities invite certain responses in audiences (see Ott and Dickinson, 2008). For visual rhetoricians, the primary question is therefore: 'How do images act rhetorically upon viewers?' (Hill and Helmers, 2004: 1). Rhetorical visual analysis is indebted to Barthes' semiotic approach, with his essay title 'Rhetoric of the Image' clearly signalling the correspondences. Influenced by Barthes' *Mythologies* in his method of analysis, visual culture scholar Matthew Rampley recognizes the culture- and time-bound nature of his own interpretations as against other possibilities, but makes the point that the key question is not which of the possible readings is 'correct', but that the photograph is interpreted 'in terms of its strategic and rhetorical functions' (Rampley, 2005: 139–40). Rampley has also noted a renewed interest in the *strategic* nature of both visual communication and representation, with interpretive analysis not only involved with aesthetic concerns, but with 'the ways in which the practices of visual culture are intertwined with mechanisms of social power and ideology' (Rampley, 2005: 135).

As with other disciplines, scholars of visual rhetoric have had to contend with the traditional treatment of images as subordinate to written texts, or as mere decoration rather than a rich intertextual site for understanding cultural and political values. Visual rhetoric considers how images can pose argumentative claims through certain persuasive cues – for example, in the way that visual elements are organized to create linkages (for instance, in how the national flag appears behind a person making an address), and in the recognition of dominant tropes, intertextual references, synecdoches and metaphors (see Foss, 2005).

The idea that pictures could posit an argument received particular attention in two special issues of *Argumentation and Advocacy*, now over 20 years ago. In their introduction to the divergent essays, David S. Birdsell and Leo Groarke (1996: 1) hoped to 'spur the development of a more adequate theory of argument which makes room for the visual', and they address 'the prevailing prejudice that visual images are in some intrinsic way arbitrary, vague and ambiguous'. Referring to essays within the special issue, they posit that it is not just pictures that can be 'vague and ambiguous', as commonly argued, but also words and sentences when they are considered out of context (Birdsell and Groarke, 1996: 2).

As an example of applied visual rhetoric, Janis L. Edwards and Carol K. Winkler (1997) examine editorial cartoons that feature appropriated, abstracted or parodied versions of the Iwo Jima flag-raising picture, based on the iconic photograph taken by Joe Rosenthal in 1945. The photograph is one of the most recognizable and distinctive images from the Second World War, with a group of soldiers raising the American flag on Mount Suribachi. Epitomizing the heroic vision of war, the photograph depicts universal, faceless soldiers symbolically taking the highest point on the island – though in reality the battle for the island was to rage on and only three of the six flag-raisers would survive. For Edwards and Winkler, it is the repetitive use of the image – they found 58 cartoons between 1988 and 1996 in US newspapers or syndicates – that is of interest: 'Images used strategically in the public sphere reflect not only beliefs, attitudes, and values of their creators, but those of the society at large' (Edwards and Winkler, 1997: 289). Just as Matthew Rampley stresses, it is the 'strategic' use of the images which is paramount to rhetorical theory. In their article, Edwards and Winkler explain how the images work as *representative form* in an attempt to explore how this particular image has gained a life of its own since its original denotation as heroic achievement, thanks to parody and abstraction: 'The image has become a discourse fragment that multiple publics appropriate for diverse purposes' (1997: 297).

Overall, visual rhetoric emphasizes the value of identifying rhetorical techniques in the detailed analysis of visual artefacts, of situating this within the contexts of their interpretation, and demonstrating the extent to which they can work to support or undermine political and cultural ideas. Useful resources here are the edited collection by Charles A. Hill and Marguerite Helmers, *Defining Visual Rhetorics* (2004), and *Visual Rhetoric: A Reader in Communication and American Culture* (2008), edited by Lester C. Olson, Cara A. Finnegan, and Diane S. Hope (see Chapters 4, 6 and 9 for rhetorical analyses of visual media used to promote European cities on an international stage, fake political advertising and a film trailer, respectively).

Visual framing analysis

Visual framing analysis has emerged over the past decade as a visually focused form of framing analysis, or frame analysis. Practitioners of framing analysis, applied here specifically to *news frames*, hold that newsmakers construct realities and narratives through the use of 'framing devices' such as word choice, symbols, metaphors and, crucially, visual images, which work to promote certain themes, and offer favoured interpretations and explanations of events (Entman, 1993; Pan and Kosicki, 1993). In earlier studies about news frames, visuals were often mentioned, but were rarely included in a systematic or detailed manner. For instance, Todd Gitlin (1980) defined media frames as *'persistent patterns of cognition, interpretation, and presentation, of selection, emphasis, and exclusion, by which symbol-handlers routinely organise discourse, whether verbal or visual'* (Gitlin, 1980: 7, emphasis in original).

Robert Entman's much cited definition focuses on how frames structure certain aspects of reality, 'to make them more salient in a communicating text in such a way as to promote a particular problem definition, causal interpretation, moral evaluation, and/or treatment recommendation' (Entman, 1993: 52). Due to the 'iconic ability' and indexical qualities of photographs, Paul Messaris and Linus Abraham (2001) argued for a central role for visual images in framing research, with images reinforcing cultural stereotypes which may not even be referred to in the lexical-verbal text. According to Messaris and Abraham, it is the perceived natural and factual properties of images that make viewers less likely to be aware of the effects of visual framing (see also Geise and Baden, 2015; Powell et al., 2018).

Visual framing analysis looks, then, to the patterns of 'selection, emphasis, and exclusion' (Gitlin, 1980) of news photographs to gain insight into the tacit cultural and ideological assumptions of the newsmakers and implied readers. Visual framing analysis is commonly applied to studies of war and conflict, or political coverage, both of political candidates and protesters, and to some degree in health communication (Parry, 2010a; Corrigall-Brown and Wilkes, 2012). Visual framing studies tend to focus on photographic images published in print, news agency libraries or online articles, but has also encompassed cartoons, documentaries and television news. More recently, visual framing has been applied to social media images – for example, to demonstrate how images on Twitter functioned to augment Israeli frames and create a shared identity during the 2014 Gaza conflict (Manor and Crilley, 2018).

As others have pointed out, there is no single agreed method of applying visual framing analysis to media texts: one key difference is whether the images are analysed by themselves or jointly with the surrounding linguistic content.

The majority of visual framing analyses employ a content analysis that enables researchers to chart salience of certain features over time, and also to conduct comparative work – for example, comparing different conflicts or election coverage. One core interest of visual framing analysis is the persistence of certain forms of portrayal, or visual tropes, and how such imagery becomes familiar or, alternatively, contested by competing frames across a diverse range of news media.

In content analytical approaches to visual framing, the primary unit of analysis is often the news photograph. As explained above in the section on content analysis as a method, the researcher creates a codebook enabling the systematic collection of relevant characteristics, which could include both compositional elements (angle, focus, distance), thematic or topic related variables – for instance, subject of the photograph, gender, age, graphic nature, symbolic features – and possibly also linguistic context and layout. The visual content is then considered in relation to how it works to frame certain people, actions and events, and how this supports the narratives, rationales and moral positioning of the adversaries.

Lulu Rodriguez and Daniela Dimitrova's (2011) four-level model of visual framing provides a semiotics-influenced approach, suitable for a qualitative study of a limited corpus and especially for televisual or video images. Clearly referencing Barthes' layers of meaning in his 'Rhetoric of the Image' essay, Rodriguez and Dimitrova's four levels of visual framing encompass (1) visuals as denotative systems, (2) visuals as stylistic-semiotic systems (conventions such as close-ups signifying intimacy), (3) visuals as connotative systems (the culture-bound ideas or concepts attached to the people, things or events depicted) and (4) visuals as ideological representations (how images are employed as instruments of power). For an example of a study applying this approach to televisual coverage, see our co-authored paper with Aleksandra Krstić on the 2010 Belgrade Pride parade (Krstić et al., 2017).

As can be noted in this final semiotics-influenced model of visual framing, the boundaries between the methods and approaches outlined here are 'leaky', but the driving concern remains the same: identifying where the explanatory and symbolic power lies in news media representations and how this is constructed through visual conventions and selective portrayals (see Chapter 8 for a case study on the visual framing of ISIS propaganda in British news).

CHAPTER CONCLUSION: BEYOND AND BACK TO THE IMAGE

Taken together, the six methodological approaches that we have just outlined place various degrees of emphasis on the site of the image, rather than on production, circulation or audiencing as such. Naturally, it is understandable and desirable that major methodological approaches to visual analysis ought to focus on the images themselves as their primary objects of study. This said, these approaches are also all compatible and, according to one's research questions, at times also require the adoption of methods aimed at gaining insight into how images are made, how and where they travel, and how they are interpreted by different groups of people.

Through several of our case studies, in this book we show the benefit of integrating research on the production or reception of images into an analysis of their meanings. For example, we use in-depth interviews with photographers to understand both the choices they make to portray their subjects in a particular way (see our case study on Everyday Africa in Chapter 4) and the ways in which specific professional practices, labour conditions and commercial imperatives contribute to shaping their images (see our interviews with photographers who sell their stock photos through Getty Images in Chapter 10). Likewise, for our social semiotic analyses of Vice Media and the Airbnb brand in Chapters 10 and 11, we draw from information found in news and trade media to understand the role of media ownership and marketing strategies in shaping the 'politics' of images. As Hansen and Machin highlight (2019: 9), through in-depth interviews with image-makers and the collection of evidence on the broader economic and social structures in which producers of visual communication are embedded, 'we can begin to find out why media content is the way it is'.

As far as reception is concerned, we used methods including questionnaires, visual elicitation techniques and focus groups to find out how and whether the emotions and judgements that ordinary viewers attach to particular images resonate with broader cultural assumptions about the semiotic 'power' or 'appeal' of those and similar images. For example, in Chapter 8, for our case study on the 'iconic' image of the Syrian boy pictured in the back of an ambulance after being rescued from an airstrike in Aleppo, we asked a group of students to answer several questions about how that image (with or without a caption) made them feel, using their own words.

In Chapter 11, our case study on the logos of a selection of digital media and technology brands was based on Q-methodology, a research method that uses visual elicitation to both compare different opinions and identify shared perspectives among a range of individuals (O'Neill et al., 2013). The case study was centred on a survey with a Q-sort exercise where participants were asked to place a number of images into different categories expressing a range of points of view regarding how 'good' or 'appealing' they thought that the design of different logos was. The survey was then followed by focus groups with a smaller sample of participants who were asked to further elaborate on their own and others' responses, thus also generating additional insights into the ideologies and overall discourses that shaped their relationship with brands and their logos. Overall, this approach contributes an evidence-based understanding of viewers' underlying attitudes and values (Lobinger and Brantner, 2015).

In addition to engaging with aspects of production and reception to understand how particular contexts and practices shape the meanings of visual communication, in this book we also ask questions about the digital circulation and recontextualization of social media images and commercial imagery from global image banks. The digital methods needed for this kind of research are not as accessible as the qualitative methods described above in relation to production and audience research, both because they 'are not currently fully developed' (Rose, 2016: 306) and often require advanced coding skills. This said, there is an increasing availability of tools developed by researchers with these skills in order to enable other researchers to run basic analyses of how images are shared and used online. To achieve these goals, these tools may use searches of the images' tags or metadata or, on the other hand, reverse searches of the images themselves (see Rogers, 2013 for a comprehensive introduction to digital methods; for applications to research on visual images, see also Rose, 2016; Highfield and Leaver, 2016). In Chapter 7, our case study on Twitter images of Black Lives Matter is centred on a sample of tweets obtained from a search of key hashtags, which we collected through the Mecodify tool that was developed by Walid Al-Saqaf for the EU-funded project Media, Conflict and Democratisation (MeCoDEM) led by Professor Katrin Voltmer at the University of Leeds. Along the same lines, for our case study on the Lean In Collection by Getty Images, we used the Google Reverse Image Scraper developed by researchers from the Digital Methods Initiative in Amsterdam to collect information on where images from this collection are shared and how they are used in a range of online media articles. It is not lost on us that the viability of these freely available tools is contingent upon the vagaries of black-box decisions made by powerful corporate search engines like Google. But this is a growing area of enquiry in visual communication research nonetheless and one where researchers are working collaboratively to find inventive though rigorous ways to 'scrape' data from proprietary platforms that would otherwise remain completely opaque.

As a whole, researching aspects of production and reception can enrich visual analysis with nuanced accounts of how both image-makers and viewers make sense of and engage with the meaning(s) of the images that they produce or consume, but only as long as the visual detail and specificity of the images under study is not lost in an analysis of the broader sociological premises and implications of production or reception as such (Rose, 2016). Likewise, gaining insight into where images travel and how they are reused and refashioned across digital platforms and media texts can enhance our understanding of how the meanings of visual images are exchanged and transformed in broader contexts of media culture. In the end, however, we need to keep in mind that to illuminate what visual communication is and does, it is not sufficient to examine aspects of meaning-making that are located outside the image. While moving beyond the image helps us ground our analyses in important observations about the real-life contexts of visual communication, we always need to go back to the image itself for a deeper engagement with its meanings and implications. For this reason, in this book we use the six overarching methodological approaches that we discussed earlier in combination with the methods that we have just outlined. It is for the same reason that our 18 empirical case studies address a varied selection of research questions by carefully integrating considerations about the textual and contextual features of visual communication, thus also demonstrating how a wide range of images 'work' in media culture.

PART I
IDENTITIES

3 Envisioning the self in digital media   37

4 Communicating visions of collective identity   61

5 Ways of seeing difference beyond stereotypes   85

3
ENVISIONING THE SELF IN DIGITAL MEDIA

The next three chapters come under our broader theme of 'identities'. The notion of identity is linked to how we see and define ourselves in relation to others and the groups to which we belong. Identity is also and foremost where the personal meets the social (Woodward, 2002). Both personal and collective identities are constrained by material and symbolic structures, and are therefore shaped by power (Foucault [1975] 1995). Our sense of who we are and where we belong is not simply the outcome of how we choose to live our lives. There is also a growing appreciation that our identities are not necessarily fixed or easily categorized into labels imposed upon us by public institutions or the mainstream media. We have multiple identities or aspects of our identity that hold varying significance to us.

In this chapter, we explore the relationship between self-mediation and visual communication, with a particular focus on social media and digital culture. As a whole, this chapter will:

- introduce research on self-representation and self-branding in digital culture;
- discuss the role of visual communication in self-mediation and its relationship with a range of social media practices, with a particular focus on selfies and Instagram;
- examine the ways in which the top social media influencer Jenna Marbles visually performs and negotiates her identity successfully on YouTube (Case study 1);
- analyse the visual resources used by trans individuals to represent themselves on Instagram in ways that may be both constraining and empowering (Case study 2).

Overall, the chapter's case studies focus on how personal identities are visually mediated, constructed and mobilized in social media. We now turn to an overview of how the 'self' is visually represented, promoted and ultimately also 'envisioned' through a range of digital media practices.

PERFORMING IDENTITY ON SOCIAL MEDIA: FROM SELF-REPRESENTATION TO SELF-BRANDING

Identities are something we invest in affectively and bodily, and which we perform on a daily basis. As philosopher Judith Butler (1990) argues in her groundbreaking book *Gender Trouble*, identity traits such as gender, for example, are constructed through the repetition of stylized acts over time; in other words, it is one's daily performance of gender that creates their gender. Overall, visual markers of identity can reflect both our own individual choices – for example, in matters of clothing or hairstyle – and the physical and social characteristics that we might be less able to change, such as age, ethnicity and, to varying extents, also gender or nationality. Research on identity in digital culture has examined the politics and potentials of **self-representation** across widely popular platforms like Facebook, Instagram and YouTube. Platforms that tend to cater to niche audiences such as subcultural groups or lifestyle communities – for example, Tumblr and Pinterest – have also been key to an understanding of how individuals perform their identities online.

Nancy Thumim (2012) defines self-representation as a genre that is premised upon the capacity to deliver authentic and individual accounts of 'ordinary people'. As a genre that pervades participatory digital culture, self-representation is now widely used by publicly funded institutions – for example, museums – and privately owned media companies alike, and naturally it is also at the very heart of social networking. Thumim highlights that the process of self-representation is different from self-presentation (Goffman, 1959), because it is always mediated and results in bounded textual objects that have 'the potential for subsequent engagement' (Thumim, 2012: 6). Self-representations can be written, visual or quantitative (as in the case of self-tracking apps for health and fitness, such as Strava or MyFitnessPal), and often these three different modalities are intertwined (Walker-Rettberg, 2014). Among others, profile pictures, family photographs and homemade videos are visual self-representations that are now ubiquitous online.

Academic literature on self-representation focuses largely on the implications of mediated accounts of the self by ordinary people for self-expression, self-empowerment and the democratization of media culture at large. Some scholars, however, emphasize that self-representation is often linked to both labour and commerce, and that, in fact, **self-branding** has become a dominant practice in participatory culture. Alison Hearn (2008: 198) defines self-branding as a form of self-representation aimed at producing 'cultural value and, potentially, material profit', which in turn also feeds the corporate interests of powerful media institutions. In addition to purposefully crafting 'branded' selves to obtain more 'likes', 'friends' or 'followers', many social media users also seek to monetize their online presence. While internet celebrities and social media influencers are clearly in the business of promoting themselves online for financial profit (Abidin, 2018), there is also an increasing number of freelancers, journalists and entrepreneurs, among others, who work incessantly on crafting appealing self-representations across various platforms to 'make it' in the culture and creative industries (Marwick, 2013).

Overall, self-representation and self-branding are clearly related, if not interdependent, in participatory digital culture, but their relationship is fraught by the perceived contradiction between 'genuine' self-expression and 'instrumental' performativity. In their book on YouTube, Burgess and Green (2018) demonstrate that self-representation and self-branding are not mutually exclusive in user-generated media production. Vlogs, video tutorials, homemade music videos, and personal videos of pets and family on this popular social media platform often reflect both the convergences and frictions between self-expression and monetization that set apart vernacular creativity on the internet. We will explore some of the ways in which the tension between self-representation and self-branding may be communicated visually in this chapter's first case study.

THE SELF(IE) IN DIGITAL CULTURE

So, what does the 'self' look like in media culture? While this is a question that ought to be tackled from multiple perspectives, there is no doubt that what we have come to know as the 'selfie' has become the most popular and widespread type of visual self-representation. In 2013, 'selfie' was announced as the *Oxford Dictionaries* Word of the Year, due to a sharp rise in online mentions of this English-language neologism. In 2016, 24 billion selfies were uploaded to Google Photos alone, and a year later users of the selfie-centric mobile photo-messaging app Snapchat sent over 3.5 billion snaps each day. Finally, as of the end of 2018, image-heavy social networking sites like Instagram and Facebook had 1 billion and over 2 billion users worldwide, respectively. Given the significance of selfies for contemporary media culture, research on selfies covers a vast range of questions and issues, including but most certainly not limited to the relationship between self-representation and both self-branding and (micro)celebrity (Marwick, 2015), and the ways in which young people, women and members of marginalized groups use selfies as a means to manage and express their identities online (boyd, 2014; Vivienne, 2017).

The Selfies Research Network (selfieresearchers.com) has been an especially important resource for the development of scholarship and teaching on selfies. Some of the network's founders and members have contributed key conceptual definitions of selfies and theoretical interventions in relation to overarching discourses on this phenomenon. For example, Theresa Senft and Nancy Baym (2015: 1589) define the selfie both as a photographic 'object' and as a 'gesture' that sends 'different messages to different individuals, communities, and audiences'. They thus argue that researchers should examine the politics, practices and potentials of selfies rather than their representational features alone. It is through this lens that Senft and Baym also intervene in debates on selfies' causes and effects. In particular, they debunk popular and pseudo-scientific discourses on selfies as expressions of narcissism and highlight the gendered nature of such criticisms, given that most selfie-takers are women and girls.

According to *Oxford Dictionaries*, 'selfie' is 'a photograph that one has taken of oneself, typically one taken with a smartphone or webcam and shared via social media'. Katrin Tiidenberg (2018) explains that the *Oxford Dictionaries*' definition highlights three major properties, which all need to be present, both simultaneously and in that order, for something to be considered a selfie. For this reason, she defines selfies as 'self-representational networked photographs'

(Tiidenberg, 2018: 21). This doesn't mean that selfies must necessarily portray one's face or even be sent or posted. Rather, what this means is that a selfie is always a self-referential 'object', as it points to the person who took the photo, and that it can be easily shared with others through digital means.

As a 'gesture', on the other hand, the selfie is also rooted in agentful, yet power-laden socio-technical practices. It is precisely on selfies' gestural quality that Paul Frosh (2019) focuses in order to engage with their implications for a more nuanced understanding of the self in digital culture. Using photography theory as a point of departure for his own theoretical work on selfies, he argues that the selfie is both corporeal and phatic, and that we need to see it as 'an instrument of mediated, embodied sociability' (Frosh, 2019: 131), rather than figural representation alone.

Overall, here we recognize that by virtue of their networked and social nature, selfies are not reducible to the status of self-contained, static and iconic photographic self-portraits. At the same time, we also purposefully choose to focus on the *visual* qualities of selfies and other forms of self-representation. In doing so, we aim to underscore the importance of visual detail for an understanding of the very embodied actions, social relations and digital connectivities that other scholars foreground in their own analyses.

From a visual standpoint, selfies and social media images more broadly come in different shapes and sizes. Sumin Zhao and Michele Zappavigna (2018) highlight that unlike other more traditional types of photography, including portraiture, selfies foreground the photographer's perspective and that, therefore, they ought to be seen as key visual means for social and political engagement. To illustrate this point, they use an example from Hillary Clinton's 2016 presidential campaign. During a campaign stop in Orlando, Florida, Clinton was photographed standing in front of a crowd of supporters, most of whom were young women and all of whom turned their back to Clinton in order to take selfies with her in the background. While many criticized these 'millennial' women as being more concerned with being seen than with Clinton's presence and political message, Zhao and Zappavigna point out that these selfies made it possible for their photographers to share their points of view with others in a public arena.

They also argue that selfies' ability to enact different types of intersubjectivity is what makes them appropriate for the visual communication of marginalized voices, particularly those of girls and women. Selfies are inherently apt to empower marginalized groups because they enable them to share and negotiate their perspectives in public – something that more traditional visual media such as press photos can't do. Zhao and Zappavigna's visual theory of selfies is thus in stark contrast with common criticisms of selfies as fundamentally narcissistic. By the same token, it is complementary to approaches that see selfies more specifically as a visual means for representing the self.

Based on their overarching framework, Zhao and Zappavigna (2018) identify four main types of selfies according to the different types of intersubjective relations that these realize. First, the *presented selfie* focuses on the photographer herself or on the photographer together with other subjects, as in the case of group selfies. Usually the photographer's face and upper body are most salient within the image but, unlike a traditional photograph, the presented or 'classic' selfie also makes the presence of the camera known to the viewer, albeit implicitly. Through visual cues like distortion and the reversed mirrored view that is typical of smartphone photography, together with the 'gestural' qualities that we described in the previous section, the presented selfie visually presents the photographer's perspective on the representation of her identity as an individual. In doing so, then, 'classic' selfies communicate the photographer's perspective on the self in ways that highlight both the intimacy and the performativity of self-representation.

While in the presented selfie the photographer herself is centred and typically takes up the largest amount of space within the frame, the *mirrored selfie* is a shot of the photographer's image in a mirror. Mirrored selfies typically include the photographer (who is usually shot from a greater distance), the phone and the mirror. Clearly, mirrored selfies are often taken and posted to communicate the photographer's perspective on her own body, particularly in relation to major milestones, achievements and points of pride. Not surprisingly, then, pregnancy, post-workout, weight loss and outfit selfies are all typically also mirrored selfies. Mirrored selfies can therefore also open up dialogue between the photographer and others about their shared experiences and feelings regarding how their bodies and appearance have changed alongside their identities – for example, as mothers-to-be or as fitness enthusiasts.

Finally, both the *inferred selfie* and the *implied selfie* do not focus on the photographer as such, but on the other subjects, objects or sceneries included in the photos. The only major difference between implied and inferred selfies is the degree of easiness with which we can recognize the photographer's perspective. In implied selfies we always see body parts, like a hand, that belong to the photographer, and which point to the presence of the photographer. This said, in this kind of selfie the hand does not represent the photographer as such, but rather her perspective on what is included in the image – for example, the coffee cup that the photographer is holding in her hand or the child whose hand she is holding. Likewise, in inferred selfies the photographer's perspective is communicated by the presence of objects, like a cup of coffee, or of a particular setting, such as a coffee shop. While in inferred selfies there is no bodily reference to the photographer as such, the presence of 'fetishized' objects and surroundings often communicates the photographer's 'way of seeing' and affiliation with particular identities and communities.

As a whole, then, selfies can be seen as a distinctive visual genre because they 'can articulate and generate perspectives in different ways' (Zhao and Zappavigna, 2018: 1749). It is also in this sense that selfies are a key visual means for expressing one's voice in networked arenas. And while selfies have practically become shorthand for self-representation on the internet, we must not forget that there are multiple ways of visualizing and indeed also envisioning the self across a range of social media platforms. This is covered in the next section.

THE VISUAL AESTHETICS OF SOCIAL MEDIA PHOTOGRAPHY

We now turn to a broader discussion of the aesthetics of Instagram, the most widely used visual social media platform. Instagram has over 1 billion users, but in spite of its mainstream status, personal photography on this platform is far from being homogeneous. In his large-scale study of over 15 million images shared on Instagram between 2012 and 2015 in 16 global cities, Lev Manovich (2017) found significant variations in the subjects and styles of photographs across locations and demographics. On Instagram, the very nature of what looks 'ordinary' changes considerably across cities, social and age groups, and even subcultures. In addition, Instagram is used by different groups of people in different cities. For example, the study found that in large world cities like New York, London and Moscow, Instagram was used widely by locals and tourists alike, whereas in other cities, like Amsterdam, for example, it was most likely used by the young English-speaking elites or by members of the culture industries, including so-called creatives and lifestyle promoters.

Ultimately, the content of Instagram images, including selfies, is highly dependent on context and their meanings are also shaped by how their subjects are photographed – for example, as part of a spontaneous snapshot rather than a styled composition. Manovich (2017: 39) therefore argues that, in analysing social media imagery, we must consider images' 'content, their aesthetics and their larger context'. Within this general framework, his research shows that we can divide Instagram images into three main categories according to their visual aesthetics. These are *casual*, *professional* and *designed* photos.

Casual photos focus heavily on what Manovich calls 'the human world' and include a large amount of selfies. He notes that these are photos that are taken and uploaded with the purpose of visually documenting and sharing an experience or portraying a person or a group of people. For this reason, the visual aesthetics of casual photos are centred on conventions originating from vernacular photography, which dictate both what ought to be photographed and how it should be shot. According to some of these conventions, for example, individuals and groups of people should be portrayed at the centre of an image, and the horizon line of landscapes should be shot horizontally rather than through strong angles. On the other hand, food and details from interiors ought to be photographed from a diagonal or otherwise 'quirky' angle. Finally, according to the conventions of vernacular photography, there are certain subjects that are worth photographing, including sunsets, historical landmarks and tourist attractions. As an example, in an analysis of 10.5 million Instagram photos shared in New York in 2014, 13.5 per cent of all images were shared in Times Square. It is in this sense that, as Manovich (2017: 53) notes, 'casual photography is anything but casual'.

On the other hand, professional photos focus on the portrayal of ideal subjects such as natural and urban landscapes, and are visually shaped by the rules of 'good photography' – for example, the 'rule of thirds', proper exposure and symmetrical composition, and colour balance. Finally, designed or 'styled' photos are grounded in the aesthetics of modern graphic design. With regard to content, these are also photos that most often 'show the designed environment, as opposed to nature' (Manovich, 2017: 69). The aesthetics of designed photos are what we typically associate with the 'Instagram look' adopted by a range of hip, creative Instagram users, often as a means for self-branding. This is also what Manovich calls 'Instagramism', the style of the 'global digital youth class' that emerged in the early 2010s and is centred on minimalist settings, compositional flatness, asymmetric arrangements, both brightness and emptiness, and a focus on moods and atmospheres.

Although we have come to ascribe the designed photos of Instagram and other social media platforms to the work of influencers and those who aspire to monetize their social media presence, Manovich claims that Instagramism is not solely related to self-branding or self-promotion. Instead, he argues, designed photos demonstrate that '*art* and *commerce*, *individual* and *corporate*, *natural* and *fabricated*, *raw* and *edited*' (Manovich, 2017: 124, emphasis in original) are increasingly merged together, and are not usually seen as being in contradiction with one another by social media users. In analysing social media photography, then, it may not be useful to focus on attempts to distinguish 'authentic' self-representation from purely 'promotional' or 'aspirational' self-branding.

In this chapter's case studies, we therefore focus on the visual resources used by social media users to communicate their identities in ways that are both personal and public, and which may therefore combine authenticity and intimacy with performativity and self-promotion. The first case study is an analysis of how the social media influencer, Jenna Marbles, visually constructs and negotiates her gender performances on YouTube in the face of widespread hostility against female internet celebrities. In a related manner, the second case study focuses on how ordinary trans individuals balance different communicative goals in Instagram photos aimed at celebrating their trans identities.

CASE STUDY 1: THE VISUAL PERFORMANCES OF JENNA MARBLES ON YOUTUBE

With over 19 million subscribers and nearly 3 billion video views as of March 2019, Jenna Marbles is one of the most successful female YouTube personalities. Jenna Marbles, whose real name is Jenna Nicole Mourey, began her career as a vlogger on YouTube in early 2010, posting comedic videos full of profanity that covered topics ranging from her dogs and make-up to relationships and sex (O'Leary, 2013). One of her very first videos, 'How to trick people into thinking you're good looking', received more than 5 million views in its first week. With over 400 weekly videos under her belt and a loyal audience of young women and teenage girls, Mourey is considered to be one of the most powerful women in online entertainment. Over less than a decade, Mourey has built an extraordinarily successful self-brand which, at the time of writing, had a net worth of US$6.5 million.

In their research on the reasons underlying Mourey's major success, Lindsey Wotanis and Laurie McMillan (2014) found that she performed her gender on YouTube in ways that enabled her to mitigate the consequences of this platform's hostile and misogynistic environment. In spite of the platform's published community guidelines, which defend free speech while also highlighting that abusive behaviours and violent or sexually explicit content are not permitted, YouTube female performers systematically receive more negative feedback. Wotanis and McMillan note that, despite her major success, even Mourey received more aggressive and sexually explicit comments than her male counterparts, but that she was also able to negotiate such hostility through a set of performance strategies that made her simultaneously appealing to different types of viewers. Likewise, Emma Maguire (2015: 73) argues that Mourey 'has developed a strategy to get around this weighted system which requires girls to perform "hot" (if they want to be shareable and consumable), but which also punishes them for it'.

This case study builds on the existing research on Jenna Marbles by focusing more specifically on the role that key visual aspects of Mourey's videos, rather than their overall strategies, play in negotiating different facets of her gender performance on YouTube. We now turn to an explanation of our research questions and approach to analysing a selection of Mourey's YouTube videos.

Research questions and approach

Our main research questions for this case study are:

- What are some of the key visual techniques used by Mourey to communicate her identity in her YouTube videos?
- How do these visual techniques contribute to constructing and negotiating Mourey's gender performances?

Here we adopt a broadly **rhetorical approach** grounded in film theory to examine Mourey's three most popular videos. These are, in order of popularity: 'How to trick people into thinking you're good looking' (posted on 9 July 2010), 'What Girls Do In The Car' (28 September 2011) and 'How To Avoid Talking To People You Don't Want To Talk To' (15 February 2011).

In this short analysis, we are especially interested in how **video editing** and key aspects of the ***mise-en-scène*** (for example, the setting and the actors, but also make-up and props) work together to engage us as viewers in ways that contribute to complicating our understanding of Mourey's gendered identity. As David Bordwell and Kristin Thompson ([2004] 2017: 294) have argued in their now classic work on cinema, 'editing strongly shapes viewers' experiences, even if they are not aware of it.' In coordinating one shot to the next, editing is not only responsible for our understanding of the film's story as such, but also, and perhaps most importantly, for how we perceive and engage with its meanings – for example, in terms of the various associations we may make between specific characters and particular moods or atmospheres. We now turn to examining Mourey's three most popular videos through the lens of Bordwell and Thompson's framework for analysing editing in film.

Findings: negotiating performances of the self through visual juxtaposition

Vloggers typically use amateur video-editing techniques to make their videos, both for practical and financial reasons and as a shared stylistic cypher for authenticity (Burgess and Green, 2018). Jump-cut editing is particularly widespread among YouTubers because it allows them to easily piece together a number of takes into a coherent whole, while also cutting out pauses and speech fillers such as 'uh' and 'oh'. Jump cuts typically link together two sequential shots portraying the same subject – in this case, the vlogger – which are taken from similar if not nearly identical camera distances and angles. Because these two shots 'are cut together but are not sufficiently different' (Bordwell and Thompson, [2004] 2017: 335) in spatial, temporal and graphic terms, there will be a visible jump on the screen, but also a more or less abrupt gap in the story. As a specific type of elliptical editing, then, jump-cut editing conveys an effect of jumping slightly forwards in time without, however, significantly deviating from the previous shot's physical location, chronological setting or visual appearance.

In turn, as a technique jump-cut editing contributes to creating the relatively fast-paced and therefore also 'witty' or 'punchy' style that is typical of many successful YouTube vloggers (Christian, 2009). Generally speaking, however, vloggers also attempt to achieve a nearly seamless feel for their videos by maintaining key aspects of the *mise-en-scène*, together with framing and point of view, unaltered. In other words, vlogs often use jump-cut editing while also approximating some of the rules of continuity editing to create compelling, incisive videos that also maintain an impression of 'liveness' and intimacy.

Instead of attempting to make her videos look and feel as seamless as possible within some of the conventions of vlogging that we have just described, Mourey plays with jump-cut editing by deliberately piecing together shots that are in contrast with one another. In doing so, she creates visual juxtapositions between different versions of herself, which also has implications for how the viewer relates to the ways in which she performs her identity on YouTube. And while these visual juxtapositions are only one of the rhetorical strategies that Mourey employs to negotiate her gender performances in the vlogging arena, we would also argue that they provide a particularly strong visual anchorage for how we experience rather than simply understand her 'self'.

Visual juxtaposition can be defined as a visual technique that relies on a more or less implicit comparison between contrasting attributes, actions, settings or moods (Aiello, 2012a). Furthermore, according to Catherine Lutz and Jane Collins (1993: 74), 'juxtaposition produces the "third effect," or new meanings evoked in viewers by seeing two photos side by side.' When it comes to film editing, an effect of visual juxtaposition can be achieved by manipulating 'graphic relations' between shots, or similarities and differences between 'the *purely* pictorial qualities of those two shots' (Bordwell and Thompson, [2004] 2017: 297). Bordwell and Thompson ([2004] 2017) also explain that key aspects of the *mise-en-scène* such as lighting, setting, costume and make-up, and the behaviour of characters and other figures in space and time can all be used as graphic elements, together with more cinematographic characteristics such as photography, framing and camera movements.

What is particularly striking about how Mourey uses visual juxtaposition is that she actively relies both on graphic continuity and discontinuity between shots. For example, in 'How to trick people into thinking you're good looking', we see her as she transforms the way she looks in front of her audience during a mock beauty tutorial. Rather than gradually applying make-up or doing her hair in front of the camera, however, Mourey chose to piece together contrasting shots that show her abrupt changes through make-up and hairstyling (Figures 3.1 and 3.2).

FIGURE 3.1 Visual juxtaposition and graphic contrast in jump-cut editing from 'How to trick people into thinking you're good looking' by Jenna Marbles. Source: YouTube.

As Maguire (2015) notes in her own analysis of this video, here Mourey carefully used jump-cut editing between verbal instructions like 'Literally cake a bunch of make-up on your face' and 'Don't forget your hoochie lipstick!' to represent herself as increasingly unrecognizable, while also visually mobilizing a stereotypical definition of 'hotness' to her own advantage.

At the same time, in Mourey's videos different shots are also carefully, though seemingly casually, coordinated to match graphically in a few important ways. In both pairs of shots reproduced in Figures 3.1 and 3.2, her transformation is dramatic, but it is limited to only one aspect of her own self-representation while the overall graphic composition of the two shots remains unaltered. The shot's framing, her posture and placement within the shot, and both the video's location and camera angle in relation to the background, together with her hair in one case (Figure 3.1) and her made-up eyes in the other (Figure 3.2) are visually identical across the two shots. In other words, there is a 'graphic match' between most of the two shots' pictorial qualities except for the one detail that Mourey wants to foreground as an abrupt change in her appearance through jump-cut editing – namely, her made-up eyes in Figure 3.1 and her hair in Figure 3.2. It is clear that Mourey's most popular video focuses on demonstrating how incremental and, by way of jump-cut editing, also sudden transformations in her appearance can turn her into someone who is not only 'good looking' but also completely different from her 'authentic' self. This said, throughout this video Mourey uses graphic relations in editing in ways that highlight differences but also similarities between her 'unattractive' and 'attractive' self.

FIGURE 3.2 Visual juxtaposition and graphic match in jump-cut editing from 'How to trick people into thinking you're good looking' by Jenna Marbles. Source: YouTube.

In her other two top videos, Mourey uses visual juxtaposition in a similar though not identical manner. In 'What Girls Do In The Car', jump-cut editing is used throughout to pattern shots where Mourey's facial expressions alternate between the pouty and wistful airs that we have come to associate with the online self-representations of female social media influencers and the exaggerated, zany grimaces that are more common among (typically male) comedy vloggers – for example, as Mourey goes from saying, 'Hey, you can see right through these pants' to 'It's my vagina!' (Figure 3.3). This combination of jump cuts and contrasting facial expressions against the backdrop of otherwise visually consistent shots contributes to enhancing Mourey's ability to present and re-present herself as both conventionally attractive and more broadly appealing.

FIGURE 3.3 Visual juxtaposition through facial expressions in jump-cut editing from 'What Girls Do In The Car' by Jenna Marbles. Source: YouTube.

In 'How To Avoid Talking To People You Don't Want To Talk To', this approach to visual juxtaposition is emphasized by means of repetition. Here, Mourey is consistently shot against the same background from an identical camera angle. However, each time that she describes a situation in which one may not want to be approached or talked to, she concludes her train of thought by saying, 'You give them one of these' or 'Just stand there', and this is immediately followed by a jump cut to a shot portraying her with the same frozen expression on her face (Figure 3.4). This happens five times during the video, thus underscoring the contrast between her implicit gender performance as a carefully groomed and made-up young woman and her explicit comedic performance as a farcical mask of sorts.

Overall, Mourey's approach to visual juxtaposition emphasizes both differences and similarities across her multiple performances, and therefore also works as a tactic 'that anticipates a hostile viewer who will dispense such insults as a way of punishing Mourey for an attention-seeking self-representation' (Maguire, 2015: 81).

FIGURE 3.4 Visual juxtaposition through repetition in jump-cut editing from 'How To Avoid Talking To People You Don't Want To Talk To' by Jenna Marbles. Source: YouTube.

Case study conclusion: balancing discontinuity within continuity in the Jenna Marbles self-brand

As a whole, Mourey uses competing or contradictory representations of herself to her advantage – for example, by performing both the role of a nerdy girl wearing a retainer and that of a hot Barbie look-alike in her most popular video (Maguire, 2015). In this case study, we have focused on visual juxtaposition through jump-cut editing by analysing graphic relations between shots to demonstrate that Mourey's seemingly casual, DIY approach to video editing is instead carefully orchestrated to foreground different facets of her identity, which, however, also co-exist under an overarching self-brand. In other words, Mourey clearly balances discontinuity within continuity, or multiplicity and coherence, in her approach to self-representation and self-branding (see Floch, [1995] 2000). In doing so, she ultimately also manages to mobilize a widely appealing self-ironic and even self-deprecating rhetoric while reinscribing traditional gender norms into her online performances (Wotanis and McMillan, 2014).

CASE STUDY 2: MAKING THE TRANS SELF VISIBLE ON INSTAGRAM

In September 2015, Laverne Cox posted a picture to Instagram of herself wearing a strappy one-piece swimsuit on Fire Island, using the hashtag #TransIsBeautiful. In an interview with *Cosmopolitan* just a few days later, the Emmy-winning actress from *Orange is the New Black* (*OITNB*) said that she had started this hashtag 'as a way to celebrate all those things about us that make us uniquely who we are' (Manning, 2015) and, specifically, to elevate trans beauty in a society where trans and gender non-conforming people are misgendered, harassed and subjected to physical violence on a daily basis.

While Cox is especially well known as an activist celebrity, over the last decade the movement to include and support trans identities in everyday public life has grown considerably across social spheres. At the same time, television shows that foreground trans characters and themes like *OITNB*, but also *Transparent* and *Pose* have reached mainstream popularity, and the demand for images of trans people among advertisers and marketers has grown so much that leading corporate image banks like Getty Images and Adobe Stock have created entire stock image collections devoted to gender fluidity (Aiello and Woodhouse, 2016; Cain Miller, 2018; see Chapter 5 for further discussion of *OITNB* and Chapter 10 for a visual analysis of stock photos).

While underscoring the importance of including LGBTQ voices in public life, both scholars and activists have pointed out that there are pitfalls to increased visibility. Following the growing 'professionalization' of LGBTQ themes, celebrities and characters in popular culture, the inclusion of these identities in marketing and mainstream media has largely resulted in the 'normalization' of their portrayals at the expense of racial, gender and sexual diversity (Sender, 2007). Mainstream media representations of trans people, in particular, have most often treated trans characters as freaks that ought to be distanced and objectified, and as deviants who are ultimately unable to 'fit in' (Horak, 2014).

Thanks to social media, however, trans visibility is no longer the almost exclusive domain of traditional mainstream media. Forms of everyday activism like digital storytelling, vlogging and hashtag activism have enabled trans activists, celebrities and ordinary individuals alike to represent themselves while also reaching wider audiences, connecting with those who share similar stances and identities, and extending the longevity of their message (Vivienne, 2016). In this case study, we examine trans self-representations posted on Instagram, the most widely used social media platform, with hashtags aimed at promoting trans visibility.

Research questions and approach

This case study is based on two key research questions:

- What are the main visual resources that individuals using key hashtags linked to trans visibility use to represent themselves on Instagram?
- What do these visual resources tell us about the kinds of trans visibility that these self-representers may or may not be able to promote on this social media platform?

Here, we use the term 'trans' as this contraction is more widely used than 'transgender' by self-representers who identify as such (Raun, 2016). It is also important to note that some scholars and activists advocate for the term 'trans∗', as the asterisk points to the possibility of multiple endings, and therefore also accommodates gender-diverse identities that may not be otherwise included in dominant definitions of 'trans' or 'transgender' (Vivienne, 2017).

To analyse the images included in this case study, we use a broadly **semiotic approach** drawing from Sumin Zhao and Michele Zappavigna's (2018) research on the visual semiotics of selfies and Lev Manovich's (2017) work on Instagram's aesthetics. In doing so, we also connect our semiotic account with broader considerations about **Instagram as a platform**, particularly in relation to its aesthetic and algorithmic constraints (Duguay, 2016; Serafinelli, 2018).

Our sample includes images with the following hashtags: #transpride (with 836,179 posts at the time of our research), #transisbeautiful (829,524 posts), #thisiswhattranslookslike (202,631 posts), #transandproud (93,657 posts). At the time of our research, these were the most widely used hashtags in relation to trans visibility on Instagram, and at times they were also used together to tag the same images. For the purposes of this short case study, we created a small sample of 36 Instagram posts. This sample included only the top posts for each of these hashtags, and this was a choice that enabled us to focus specifically on images that were made most visible by the platform itself. In order to protect the anonymity of the individuals who posted and appeared in these images, we decided not to include links to the images in our sample here. Following Zhao and Zappavigna (2018), we also made the individuals portrayed in the images that we present in this case study unrecognizable by modifying these images through the Prisma photo-editing app's 'Curly Hair' sketch filter, and by concealing the eyes of subjects whose faces were clearly shown in the images.

Findings: trans self-representation between the ordinary and the extraordinary

Our sample of the 36 top Instagram posts across the four hashtags listed above resulted in 32 unique images. Four images appeared in the top posts for two hashtags each – for example, a trans woman's selfie was featured both under #transisbeautiful and #thisiswhattranslookslike, while a post-surgery portrait of a trans man appeared under #transpride as well as #transisbeautiful. It was also immediately obvious that the sample was heavily skewed towards the representation of male to female transition, with 24 out of the 32 unique images in our sample portraying trans women. Across these 32 unique images, we identified four main ways of communicating one's trans identity visually. These are, in order of prominence: presenting, confirming, affirming and contextualizing the trans self. Some of the images fit into more than one of these four themes. We now turn to a descriptive and interpretive analysis of the visual resources that set apart each of our four key themes (Thurlow and Aiello, 2007; see Chapter 2).

Presenting the trans self through classic and mirrored selfies

Perhaps not surprisingly, well over a third of the images in our sample (13) were selfies. 'Classic' selfies, which Zhao and Zappavigna (2018) define as 'presented selfies', were dominant, but there were also two mirrored selfies. One of the two mirrored selfies appeared among the top posts for two hashtags (#transpride and #transandproud) and the other one was also a post-surgery self-portrait of a trans woman wearing lingerie. All of these selfies portrayed young white trans women who, on the whole, were also conventionally attractive and conventionally feminine. For example, most of them were slim, had long styled hair and overall also 'passed' as cisgender women – that is, women whose gender identities correspond with the sex that was assigned to them at birth.

It is also important to note that the vast majority of these selfies were fairly 'casual', as they had been shot in everyday settings like bathrooms and bedrooms, and even when the background was indiscernible, natural lighting and minimal make-up were prevalent (Manovich, 2017). Where filters had most obviously been used, this was usually done to soften the overall look of the image rather than create dramatic effects. Most of the classic selfies in the sample made the portrayed subjects' eyes and facial features most salient, instead of highlighting more overtly sexualized features like their chest or lips. With the exception of the post-surgery mirrored selfie mentioned earlier, all of these trans women were also dressed fairly conservatively, with classic T-shirts and tops that showed little or no cleavage being prevalent (Figure 3.5). As a whole, across these top Instagram posts the trans (female) self was presented as conforming to conventional beauty standards, while also being approachable and ordinary.

FIGURE 3.5 Conventional beauty and ordinariness in a selfie by a trans woman. Source: Instagram.

Confirming the trans self through post-surgery and before and after images

The second most prominent type of images in our sample were post-surgery and before and after images (ten overall). In contrast with the selfies that we have just examined, these images contributed to 'confirming' rather than simply 'presenting' the trans self. This is because they provided evidence that transition had occurred; though in different ways, they also communicated that the desired result had been obtained.

Before and after images engage the viewer in a comparison between an 'old' and a 'new' self through visual juxtaposition (Vivienne, 2017). Five out of six of the before and after images in our sample documented male to female transition, and were centred on strong visual contrast between a conventionally and even stereotypically past male self and the Instagram user's female self. 'Before' images portrayed bearded, suited up, muscular and both short-haired and bare-chested men, whereas 'after' images were a combination of classic and mirrored selfies that showed more than just the subject's face, thus foregrounding a combination of physical traits in addition to facial features, including hairstyles, clothing and body shape (Figure 3.6). Before and after images also connected the 'old' self to the 'new' self visually by means of similar poses, frames and settings. For example, in one case both the 'before' and 'after' images were medium shots cropped at the subject's thigh, and in another case both images were selfies taken in the car while wearing a seat belt.

FIGURE 3.6   Visual contrast in a trans woman's before and after image. Source: Instagram.

The four post-surgery images in our sample were more diverse from a visual standpoint. A posed portrait of a trans man wearing a dark skirt over dark trousers and with his shirt off was perhaps the most celebratory among these, as this was also a full-length posed frontal portrait that centred and emphasized the subject's body shape and the scars on his chest against the darker background of a wooded landscape topped by a bright blue sky.

Taken together, all of these images provided these Instagram users' perspectives on trans subjectivity as being both teleological and embodied. Here, the trans self was portrayed both as the final and perfect outcome of a project 'directed toward the end of living full time in the desired gender' (Horak, 2014: 580) and, particularly in the case of post-surgery images, also as a process requiring major forms of intervention on the body. In doing so, these images offered visual evidence of such transformations, confirming that the bodily and life changes necessary for a fulfilled trans life had taken place. By confirming that the self-representer was trans, the images also and foremost 'coded' the trans self both as an extraordinary achievement that may even look magical (Horak, 2014) and as a physically demanding project that is, however, also worthy of being undertaken and celebrated.

Affirming the trans self through posed portraits

Seven images in our sample were what Manovich (2017) would define as 'professional' photographs – or medium to full-length shots portraying an idealized subject through the rules of 'good photography' – of the trans individuals (mostly trans women) who posted them as self-representations. These images were obviously taken by 'proper' photographers, portrayed their subjects in flattering poses and outfits, and in two cases they were also part of larger photoshoots that were shared as Instagram galleries. Not unlike the selfies that we examined earlier, five images out of seven portray conventionally attractive, conventionally feminine

FIGURE 3.7 Posed portrait of trans woman in a professional photo shoot. Source: Instagram.

trans women. Overall, these posed portraits 'offer' their subjects to the viewer in ways that are reminiscent of how models and actors are pictured in the official portfolios that they use for castings and publicity (Figure 3.7). Therefore, these are also images that focus largely on individuals' physical attributes rather than their actions or contexts (Aiello and Woodhouse, 2016). In doing so, they work to 'affirm' the trans self as the carrier of a beautiful exterior that is worthy of being seen and promoted.

Contextualizing the trans self through images with others

Only three images in our sample portrayed trans self-representers in the company of others and all of them included only another person in the picture. Although this was a negligible amount of images, we found that they were particularly important because, unlike other images in the sample, they contributed to contextualizing the perspectives of these individuals on their trans selves, particularly with regard to their relationships with family members and significant others.

For example, one of these images was a posed portrait that showed a young man and an older man who looked alike and stood side by side, wearing the same type of suit while striking a symmetrical pose. This carefully composed 'father and son' picture pointed to a family likeness and closeness (Kress and van Leeuwen, 2006) that proudly highlighted the young trans man's identity as part of an existing yet changing social fabric rather than as an individual if not heroic journey of the self leading to a 'break' with the past (Figure 3.8). In a related manner, the other two images were selfies that portrayed the self-representers' relationships with individuals who were meaningful to them, such as in the case of a trans woman who took a picture of herself together with the woman who was her wife when she lived as a man and who decided to stay with her through her transition.

FIGURE 3.8 Family likeness in a portrait of a trans man with his father. Source: Instagram.

Taken together, the four key analytical themes that we have just laid out show that the Instagram images in our sample represented their subjects as both ordinary and extraordinary. On one hand, the majority of these images, mostly selfies, were taken in everyday settings, and in a few cases also in the company of partners and family members. It is in this sense that the trans self was both 'presented' and 'contextualized' as ordinary. On the other hand, there was also a substantial amount of images that 'confirmed' and 'affirmed' the trans self through visual resources highlighting the extraordinary nature of the physical transformations that had taken place, such as in the case of before and after images, together with the beauty that resulted from these changes – something that was typically communicated through posed portraits and professional photo shoots. Our sample as a whole was heavily centred on images of young white individuals, with a particular emphasis on portrayals of conventionally attractive trans women who could also 'pass'.

Case study conclusion: trans visibility on Instagram as constrained empowerment

Research shows that social media self-representations contribute greatly to the development of solidarity, support and community among trans people (Raun, 2016), have the potential to promote counter-discourses about trans identities across a variety of publics (Duguay, 2016), and may even offer lifesaving narratives to particularly vulnerable individuals such as youth (Horak, 2014).

At the same time, the nature of these self-representations most often depends on the affordances and constraints of different social media platforms like YouTube, Tumblr and Instagram, together with dominant discourses centred on binary gender norms, conventional beauty standards and passing as key ideals for trans selfhood (Vivienne, 2018). This short case study has brought to light some of these tensions in relation to trans self-representations on Instagram.

The dominant visual characteristics of our sample are in line with the affordances and constraints of Instagram as a social media platform. In her research on trans and queer selfies, Stephanie Duguay (2016: 6) found that Instagram's rhetoric and socio-technical characteristics encourage the production and circulation of self-representations 'that avoid offending through assimilation with mainstream discourses' founded in conventional beauty standards, binary gender norms and celebrity-driven viewership.

This said, on a closer look, these images did more than simply conform to the individualistic ethos, inspirational discourses and aspirational aesthetics for which Instagram is known. First, even if these were the minority, a few of these top images did contribute to contextualizing the trans self as being both connected to significant others and supported by meaningful relationships. Second, when looking at what trans women wrote next to their selfies or posed portraits, it became obvious that they often posted conventionally attractive images of themselves to connect with others and lift their own or others' spirits, as can be inferred from messages like 'Been feeling pretty low on self-esteem lately so maybe this will help a bit?' and 'Love yourself, forgive yourself, and don't apologize for who you are'.

Likewise, before and after images were often paired with reflections on the consequences of gender dysphoria and feelings of inauthenticity for mental health. Finally, post-surgery self-representations were also more visually nuanced than we would expect from the kinds of images that Instagram would make most visible, as they pointed both to some of the liminal and non-binary dimensions of trans identity and to the possibility that hardship and success may co-exist throughout the transition process. For example, a mirrored selfie portrayed a conventionally attractive trans woman with a colostomy bag under her sexy lingerie, and therefore also used visual juxtaposition to highlight the possibility of being glamorous while living with some of the potential complications of gender reassignment surgery.

Social media self-representations of this kind ought to be seen as instruments for mediated and embodied sociability (Frosh, 2019) which are also 'simultaneously affirming and compliant' (Vivienne, 2018: 138), precisely because self-identity is bound up with attempts to relate and belong with 'affective publics that simultaneously constrain and reward' (Vivienne, 2018: 138; see also, Papacharissi, 2014). In other words, self-representation is neither only empowering and affirming, nor simply constrained and compliant. For this reason, it may be best to think of trans visibility on Instagram as constrained empowerment. Our sample of top images was set apart by mainstream representational and aesthetic choices that were regulated by Instagram as a platform and, more generally, by social media photography as an overarching genre. Not surprisingly, then, these images failed to portray a diverse range of trans identities. Overall, however, they also highlighted some of the issues and uncertainties associated with transitioning while focusing on celebrating trans people's beauty and successes. By portraying trans people as both ordinary and extraordinary, these Instagram posts also managed to offer somewhat more complex and empowering narratives about the trans self for multiple, networked audiences.

CHAPTER CONCLUSION: MAKING MEANING OF THE VISUAL SELF

The rise of social media has put identity and the self squarely at the centre of media culture. Maintaining an online presence through multiple social media profiles has become an everyday practice if not an imperative for many. Not surprisingly, there is now a vast amount of research on the implications of social media for how we relate to ourselves and others via user-generated content, particularly imagery. In this chapter, and across our two case studies, we have focused on two main points:

- Social media users represent themselves visually in different ways (both in terms of the contents and styles of their images) according to the particular audiences that are drawn to different platforms, together with their distinctive content generation tools and levels of moderation.
- At the same time, forms of self-representation, like selfies, YouTube vlogs or Instagram feeds, all rely on particular visual formats and conventions that constrain but also enable self-expression and both social and political engagement.

Being able to recognize oneself as fairly represented in the media clearly remains a contentious issue. There are pleasures to be gained in crafting our identities and in identifying with others because of their shared 'look', but this is balanced with 'subject-positions' constructed in regimes of representation (Hall, 1997). It is in this sense, then, that our discussion of the relationship between visual communication and identities in media culture also contributes to an understanding of this book's two other major themes – namely, 'politics' and 'commodities'. Identities may be seen as constraining or strategic in political and corporate communication alike, depending on what the intended messages and audiences are. At the same time, the visual mediation and promotion of personal and collective identities is often linked to political processes and commercial imperatives. Likewise, key aspects of social and cultural difference such as race, gender, ethnicity or nationality are increasingly used as identity resources in the mediated arenas of activism and campaigning on one hand, and both advertising and branding on the other hand.

POTENTIAL FURTHER RESEARCH

As we have seen in this chapter, Instagram's navigational features, content moderation practices and content generation tools, together with the app's official descriptions and example images (Duguay, 2016), all work together to create a self-representational formula that promotes a competitive aesthetic based on what Lev Manovich (2017) defines as 'designed photography' or even 'Instagramism'. This said, through his large-scale research on Instagram, Manovich also found that the majority of images posted on this platform are in fact 'casual' photos, but that these may be overall less visible on this platform than what we have come to consider as 'typical' Instagram images.

It is therefore also important to explore the ways in which ordinary people make sense of their own and others' visual social media practices and outputs (see Serafinelli, 2018). Key questions here could be: How do social media users make meaning of the images posted by those they follow on a particular platform? What are the connotations and overall meanings that they attribute to the visual choices made by other social media users on the same platform? This kind of study would therefore also entail interviews with research participants in combination with an analysis of social media images.

In addition, most often people use multiple social media platforms. For example, they may use Instagram to promote themselves publicly and Tumblr to connect with others about particular aspects of their identity in a more intimate manner. Another way to explore the role of visual formats and conventions in digital self-representation, then, would be to use a combination of visual analysis and interviews to understand how micro-celebrities or ordinary people use different social media platforms to represent themselves for different reasons and to multiple, though often also overlapping, audiences.

FURTHER READING

Both Thumim (2012) and Walker-Rettberg (2014) are important sources on self-representation in digital culture, and both include in-depth discussions of key visual features of particular types of self-representation. For research that focuses more specifically on the visual and aesthetic dimensions of social media photography, key sources include Manovich (2017), Zappavigna (2016), and Zhao and Zappavigna (2018).

4
COMMUNICATING VISIONS OF COLLECTIVE IDENTITY

In addition to their role in representing and mediating the self, images are also central to the formation and maintenance of collective identities. Shared narratives and symbols are key to fostering a sense of belonging among members of the same cultural or social group. In this chapter, we explore the role of visual imagery in representing and shaping national, post-national, and regional or continental identities. In sum, this chapter will:

- introduce key concepts and research on the relationship between visual images and collective identities;
- discuss the growing importance of images for the communication of particular visions and versions of identity both within and between groups;
- show how photographers use particular storytelling devices to represent a continent like Africa from an insider's perspective, with a focus on the Instagram-based project Everyday Africa (Case study 1);
- examine the main rhetorical strategies used to communicate the 'Europeanness' of cities in promotional media, specifically in relation to the European Capital of Culture programme (Case study 2).

The chapter's case studies focus on how collective identities are communicated visually for the global stage. This focus allows us to explore how countries, regions or continents, together with seemingly less tangible supranational actors like the European Union (EU from hereon), use the media to spread visual narratives aimed at promoting their identities across borders. We now turn to a focused discussion of the key concept of 'imagined community' in relation to iconic and humanist photography, together with a discussion of the ways in which collective identities are imagined and promoted in a variety of visual media.

PUTTING THE IMAGE IN IMAGINED COMMUNITIES

Benedict Anderson famously defined nations as 'imagined political communities' and stated that 'communities are to be distinguished, not by their falsity/genuineness, but by the style in which they are imagined' (Anderson, 1983: 6). Collective identities are socially constructed, and this means that they do not originate from the intrinsic characteristics of those who belong to a given cultural or social group, or what may be seen as its 'essence' (Hall, 1996). In *Imagined Communities*, the groundbreaking book in which he made this argument, Anderson stated that people who have never met and will never know each other feel that they belong together, and that this collective 'we' comes into being through shared symbols and narratives – like, for example, national anthems and flags, but also founding myths and allegedly ancient traditions (see Hobsbawm, 1983). For such sense of communion to develop, members of the same imagined community ought to access and exchange these shared symbols and narratives daily, or at least on a regular basis – for example, during celebrations and through collective rituals.

According to Anderson, this is also why nationhood emerged as a dominant collective identity framework in the nineteenth century. This is when print media, particularly daily newspapers, started being produced and circulated on a mass scale thanks to technological innovations such as the mechanization of printing and the introduction of news wire services. Consequently, individuals with similar linguistic and literacy skills but who were otherwise divided by geography and status began identifying systematically with the same events and stories, together with the more or less implicit values that these carried with them. Print media were key to the selection and propagation of narratives that 'fixed' nationhood as a unified and coherent form of collective identity defined by shared linguistic and more broadly cultural traits. And even in the face of internal diversity and fragmentation, the notion of national identity became equated with the nation-state construct almost universally.

Just like language, visual images are a key site for the negotiation, consolidation and naturalization of the major cultural narratives and social norms that define collective identities (Hall, 1997). The concept of imagined community has been widely applied to collective identities ranging from nations and regional polities to ethnic minorities, diasporic communities and even social movements. Here, we focus mainly on national and both post-national and continental identities, even though we recognize that this is a limited view on the notion of collective identity. This specific focus is due both to pragmatic reasons and to an intention to address some of the major issues of our times, particularly in the wake of the intensification of parochialisms and nationalisms across the world.

IMAG(IN)ING COLLECTIVE IDENTITY IN ICONIC AND HUMANIST PHOTOGRAPHY

In the twentieth century, photography came to the fore as a major means for the 'making' of collective identities. As Robert Hariman and John Lucaites elucidate, **iconic photographs** are especially significant in this regard, as they are 'accessible, undemanding images suited to mass-mediated collective memory' (Hariman and Lucaites, 2007a: 2). Typically, iconic photographs are widely recognized and even recalled by people of all ages and social backgrounds. However, individuals most often recognize photographs only if they are able to connect them with specific events that have 'historical and emotional significance for them and their national community' (Cohen et al., 2018: 473). As a whole, iconic photographs are images that have universal underpinnings, insofar as they are capable of rhetorically linking singular experiences with collective values and ideals. Precisely because they are significant by virtue of being linked to specific events while also having much broader symbolic meanings, iconic photographs are few and far between.

In post-war societies, some of the photographs that immortalized both the traumatic and triumphal events of the Second World War became visual commonplaces that expressed more generalized attitudes towards the nation and the values that animated it. For example, photographs like Joe Rosenthal's 'Raising the Flag on Iwo Jima' and Alfred Eisenstaedt's 'V-J Day Kiss in Times Square', both of which were taken in 1945, became iconic of the United States's victory over Japan. Both photographs also became associated with broader meanings of national pride rooted in values like egalitarianism, unity and civic republicanism on one hand, and both liberty and optimism on the other (Hariman and Lucaites 2007a and 2007b). It is therefore not an accident that, more than half a century later, Thomas E. Franklin's photograph of three firefighters raising the American flag at Ground Zero became one of 9/11's most iconic photographs by visually evoking the Iwo Jima image and affirming its ethos of patriotic citizenship.

The significance of iconic photography for nation-building in the twentieth century is also to be linked to the rise and commercial success of **humanist photography**, which dates back to the introduction of Leica's miniature camera (now simply known as 35mm) and its impact on photographers' newfound ability to shoot anywhere, both quickly and unobtrusively. After the war, particularly in Europe and the United States, humanism became the dominant style of reportage photography for a mass magazine audience thanks to its focus on the everyday lives of ordinary people through themes like 'family, community, comradeship, love, childhood, popular pleasures' (Hamilton, 1997: 94). Reflecting both anti-war and anti-fascist sentiments, humanist photography was set apart by an impulse to portray the most disparate individuals and social groups as belonging in the same greater whole, thus also affirming 'the idea of a universal underlying human nature' (Lutz and Collins, 1993: 277).

The French photographer Henri Cartier-Bresson was a pioneer of humanist photography. He developed a style based on the concept of the 'decisive moment', which he described as the photographer's ability to both recognize the significance of an event within a fraction of a second and compose the photograph in a way that gives that fugitive event its proper visual expression. Cartier-Bresson was one of the founding members of Magnum Photos, together with Robert Capa and others. Founded in 1947, just after the end of the war, Magnum is also the agency that later became equated with photojournalism par excellence (see Aiello, 2012a). Magnum photographers have documented sombre events, including wars, famines and disasters for the best part of a century. However, the agency's trademark visual style has historically relied on techniques like backgrounding, contextualization, and flattering lighting and composition that convey a sense of sympathetic identification with portrayed subjects and, overall, also a celebration of human life.

For his project *The Europeans*, Cartier-Bresson travelled across war-torn Europe to portray the 'unquestionable family likeness' (Clair, 1998: 6) of the landscapes and the people belonging to the countries that had been at war against each other, using a humanist lens to highlight the shared European identity of the continent's dispossessed and working classes. Along the same lines, Robert Capa portrayed soldiers in the battlefield in a sympathetic manner, emphasizing their shared fraternity. Furthermore, in 1948 Capa and other Magnum photographers carried out the humanist project 'People Are People the World Over' for the illustrated magazine *Ladies' Home Journal*, which showed families from different countries as they went about their daily lives. 'People Are People the World Over' was a major precursor of 'The Family of Man', the world-known humanist exhibition that was curated by Edward Steichen for the New York Museum of Modern Art in 1955 and which brought together photographs from 68 countries to represent the universality of human experience.

Both iconic and humanist photography contributed to defining and redefining collective identities in the twentieth century, particularly in the wake of post-war development and reconstruction. On one hand, and also thanks to the cheap reproduction and mass circulation of images that set apart Western mass culture, iconic photography contributed to building collective memory both through shared emotional connection with specific events and universal symbolic appeals linked to broader communal values and ideals. On the other hand, by embracing values like equality, solidarity and communality, humanist photography promoted an inclusionary vision of shared belonging which was imagined as transcending differences in cultural background and social status – a vision that was furthermore reinforced by the fact that camera-produced images were widely considered to be objective and therefore also legitimate representations of everyday life (Hamilton, 1997).

As a whole, both iconicity and humanism still animate photojournalism and documentary photography to this day, although both have also been redefined and displaced by changes in technology and visual culture more broadly. We now turn to a discussion of other types of imagery that are commonly used to communicate and promote collective identities across borders.

VISUALIZING NATIONAL AND POST-NATIONAL IDENTITIES FOR THE GLOBAL STAGE

It is not only photography that matters when it comes to the construction of collective identities through visual narratives. From breaking news images on television, like those of the burning Twin Towers on 9/11, to media representations of nationhood in events like the Olympic Games, our feelings of affinity with others and our sense of belonging in a group are shaped by multiple, often overlapping visual media narratives (Thompson, 1995).

As a whole, visual images are part and parcel of what Billig (1995) has defined as 'banal nationalism' and other forms of collective identification grounded in everyday symbols like those found in flags, currency and the media (Skey, 2011). **Promotional communication** related to travel, tourism and the branding of both national and post-national identities is also increasingly key to the construction and communication of collective identities, particularly for the global stage. In this regard, national airlines are key to the promotion of an image that ought to balance both national pride and global aspirations, and most countries spend vast amounts of money on maintaining and branding their flag carriers, whether or not they can truly afford to do so. For instance, in 2005 the Thai Airways logo was redesigned by replacing an earlier visual stereotype (a Thai classical dancer) with an orchid, or yet another self-exoticizing national emblem that lent itself to being streamlined and stylized so as to achieve a more 'international' look and feel. In a similar vein, British Airways's current corporate identity was the result of branding 'mistakes' that led to the realization that the flag carrier could not do without the Union Jack, as the British flag was a key visual symbol for maintaining trust both among domestic and global customers (Thurlow and Aiello, 2007).

Along the same lines, tourism has become central to the ways in which countries and regions represent and even engineer their identities for national and global audiences alike. Australia and New Zealand have been pioneers in promoting themselves as global destinations. Between the 1980s and 1990s, the Australian government endeavoured to design 'a positive national stereotype' (Ryan, 1990: 135) to attract desirable immigrants and foreign investors in addition to tourists. Likewise, since the late 1990s, Tourism New Zealand have used their '100% Pure' marketing strategy to promote the island country as a mythical land, untouched by the ills of modernity, and a harmonious patchwork of European, Māori, Asian and Pacific Island cultures (Shore, 2017). Overall, iconography focusing on the portrayal of happy, pristine and authentic scenarios was key to the success of both countries.

In a parallel fashion, **nation branding** has become a multimillion industry that caters particularly to the development of 'younger' countries and newer economies into appealing global players both in the eyes of their own citizens and foreign publics such as investors and tourists (Aronczyk, 2013). Visual communication is central to nation-branding consultants' attempts to communicate a country's identity as both coherent and novel. Needless to say, this is an endeavour that requires the strategic inclusion and exclusion of specific narratives. For example, the Arabic font of the logo that was created for Qatar's nation-branding campaign in the early 2000s was designed to be 'calculatedly cosmopolitan and selectively pluralistic' (Mattern, 2008: 494). This was a communicative attempt to reinforce Qatar's historical and cultural ties with other Arab countries while simultaneously demonstrating a commitment to its Western allies. Along the same lines, around the time of their 2007 entrance into the EU, post-communist countries like Romania and Bulgaria devised nation-branding schemes that relied on a bricolage of images pointing both to their pre-communist historical heritage and their links to the West (Kaneva and Popescu, 2011). By constructing a more globally appealing identity premised upon both countries' simultaneous claims of authenticity and modernity, these nation-branding campaigns also left out more complex narratives of national identity.

Both iconography and branding have also been key to the development of the EU. Clearly, the EU is not an imagined community in its own right, especially given its ongoing inability to attract widespread consensus among its citizens. As a **post-national identity**, the notion of 'Europeanness' encapsulated in the EU is narrated as a flexible combination of cultural diversity and shared values such as peace, democracy and freedom. This is something that has been promoted through various symbols, including the official motto 'United in Diversity', Beethoven's *Ode to Joy* as the EU's official anthem, the EU flag with its 12 stars, and the Euro banknotes with their combination of generic European architectural styles and abstract bridges and doors representing values like openness and a will to 'build' connections across the diverse facets of Europe (Fornäs, 2012; Sassatelli, 2017).

In addition to these official forms of iconography, visual branding has become central to the EU's efforts to promote itself as friendly, inclusive and ultimately also appealing both to its citizens and globally. As early as 2001, celebrity architect Rem Koolhaas was engaged by the European Commission to design a visual concept aimed at communicating the EU in a more attractive way. Koolhaas's visual concept merged the flags of the then 15 EU member states into a single symbol resembling a multicolour commercial barcode that could be expanded to include new countries. Due to its commercial vibe, the flag barcode was mostly a playful and provocative concept. However, it also proved to be fairly popular, to the extent that it was adopted by the Austrian Presidency of the EU Council in 2006 (Aiello, 2007). Since 2016, with the ensuing debates stirred by Brexit in the UK together with the rising tide of far-right nationalism across Europe, the EU's activities in matters of public communication have once again intensified. Together with other artists and intellectuals, for example, Koolhaas founded Eurolab, a think tank that works on devising creative ideas to rebrand the EU. Overall, communication and visual branding in particular are seen as important instruments for the deliberate though fraught construction of 'Europe' as an imagined community in the making (Aiello, 2012b).

This chapter's first case study examines the role of photography in the construction of collective identities through a producer-led analysis of the Instagram-based Everyday Africa project, which was founded by two American journalists, Peter DiCampo and Austin Merrill, with the explicit aim to debunk widespread media stereotypes about Africa. In the second case study, we use a rhetorical lens to examine the promotional media produced by cities competing for the title of European Capital of Culture, specifically in relation to the visual claims made by these cities to simultaneously promote themselves as quintessentially European, distinctively local and globally appealing.

CASE STUDY 1: A PRODUCER-LED ANALYSIS OF EVERYDAY AFRICA

Everyday Africa is a photography project that was launched in March 2012 by the American photojournalist Peter DiCampo and writer Austin Merrill to change people's perceptions of Africa by showing everyday life across the continent. The project was conceived as a collection of images entirely shot on mobile phones. The origins of Everyday Africa are to be found in DiCampo and Merrill's frustration with widespread approaches to covering Africa as a continent, through dominant media frames like war, exploitation, poverty, disease and wildlife. The two journalists had received a grant from the Pulitzer Center on Crisis Reporting to do a story about the aftermath of the post-election conflict in Ivory Coast. While speaking to both refugees and soldiers on how the country was recovering and whether violence would resume, Merrill and DiCampo realized that, even in the wake of the civil war, the daily lives of most people in Ivory Coast were fairly normal. For this reason, they also began photographing moments of daily life through their mobile phones (Dotschkal, 2014). And because they were aware that in traditional photojournalism there was little room for less stereotypical and more inclusive portrayals of life on the continent, they set out to share their photographs with others through a Tumblr blog called 'Everyday Africa'. The project was then moved to Instagram.

At the time of our research, Everyday Africa had nearly 400,000 followers, a figure that had more than doubled since 2015. Also, the number of contributing photographers, all of whom lived or had lived on the continent, had increased to over 35. Countries like Ghana, Kenya, South Africa, Nigeria and Uganda have been among the most represented, but the project has covered the vast majority of African countries. In addition, most of the project's regular contributors are African photographers including, among others, the Ugandan photojournalist Edward Echwalu, whose work has been published in global media outlets like *The New York Times* and *The Guardian*, the Ghanaian photographer Nana Kofi Acquah, who was the first recipient of the Tim Hetherington and World Press Photo Foundation Fellowship in 2016, and Sarah Waiswa, a documentary and portrait photographer based in Kenya. Increasingly, the project also features keen amateur African photographers and, at the time of writing, DiCampo had just released the first call from Everyday Africa for applications to further expand the project's pool of contributors.

Research questions and approach

In this case study, we address two main research questions:

- What are the key storytelling devices of Everyday Africa?
- How do these devices work together to represent Africa from an insider's perspective?

Here we propose a **producer-led analysis** of Everyday Africa combining information from media interviews and online materials about Everyday Africa with a research interview that we conducted with Peter DiCampo in order to understand how this project agentfully uses specific visual devices to 'repicture' Africa in a broader, more inclusive way. It is important to highlight that the main aim of our analysis is to understand how Everyday Africa uses particular storytelling devices to represent the continent from within, or from an insider's perspective, rather than examine how the project debunks particular stereotypes (we cover the representation of difference and stereotypes in Chapter 5).

Findings: mundanity and plurality to repicture Africa from within

Everyday Africa is set apart by two distinct visual storytelling devices: these are mundanity and plurality. Taken together, these devices contribute to repicturing Africa as both familiar and diverse. We now turn to examining each of these two key storytelling devices.

Mundanity

Two months after launching the project, DiCampo wrote a story on Everyday Africa for the Pulitzer Center's website. There, he stated: 'Africa can be the place of extremes that we in the West see so often. Inundated with images of incredible poverty, we also occasionally see vast wealth. But Africa can be familiar. It can also – thankfully – be boring' (DiCampo, 2012). DiCampo took the first picture of Everyday Africa inside an elevator in a government building in Abidjan, Ivory Coast's economic capital. It portrays a man with a row of lights above his head, together with other men around him and their reflections in a mirror. In the same story, he stated: 'The picture is interesting in its mundane-ness, and therein lies the truth' (DiCampo, 2012).

Everyday Africa's main storytelling device is what we define as 'mundanity', or an emphasis on the quotidian and the anecdotal rather than the extraordinary and the spectacular (see Gregg, 2004). Instead of privileging the dramatic, arresting images that are still predominant in photojournalism, Everyday Africa focuses on unremarkable moments. As DiCampo told us, the typical Everyday Africa photograph is an image of 'a simple moment made beautiful, the everyday

FIGURE 4.1 An image from Everyday Africa portraying a father and his child looking at a solar eclipse. Source: Nana Kofi Acquah for Everyday Africa. Reproduced with kind permission.

made beautiful'. For example, the image shown in Figure 4.1 shows a father and his child looking at a solar eclipse through disposable protective spectacles. This photo shows a more casual moment that, as DiCampo stated in an interview with *The New York Times*, is part of 'a general stream of daily life' (Estrin, 2012). The photograph was also part of a slideshow that Austin Merrill used in his lectures on Everyday Africa. In a lecture he gave as part of the Fillbrandt Forum at South Dakota State University, Merrill echoed DiCampo's words by stating that any of the photographs he showed as part of his slideshow could have been taken almost anywhere (Yaeger Media Center, 2015).

While focusing on mundane moments may seem like an obvious choice for an Instagram-based project, we must keep in mind that when Everyday Africa was launched, there was little to no room to portray life in Africa for a global audience as anything other than difficult or downright tragic. Instagram had only been around for two years, and at that moment in time photojournalism was still very much about shooting stories to pitch them to magazines and other media outlets. DiCampo had been living in Ghana, first through the Peace Corps, then shooting mostly for NGOs in Accra while also working on his own stories. He said: '[I was] feeling like I was putting myself back in a box over and over again of covering stories in a similar way and just often being asked to shoot similar things. I was doing a lot on disease eradication.' At the same time, DiCampo felt that photojournalism's conventional storytelling devices had become increasingly inadequate to picture 'Africa' from within rather than as an outsider, as he was 'struggling with the photojournalist impulse to make something look very severe' while also recognizing that the people he encountered were not as miserable as his photographs often showed them to be.

DiCampo's words point to a perceived need to reverse the dominant narrative about Africa as being constantly ravaged by wars, poverty and disease. In recent years, mainstream media attempts to redress stereotypical representations of Africa have ostensibly focused on promoting a positive outlook on the continent by portraying it as modern, economically attractive, welcoming and progressive

(Nothias, 2014). What DiCampo and Merrill set out to do, however, was to show Africa not so much in a positive light, or as happy and carefree (Jacobs, 2016), but rather as a 'place' like any other and, as we will explain in a moment, also as a multifaceted reality.

We asked DiCampo to tell us about a particularly representative image for Everyday Africa, and he showed us a photograph he shot in 2014 of children playing in a swimming pool in a luxury hotel in Grand-Bassam, a popular tourist resort in Ivory Coast (Figure 4.2). He pointed out that 'importantly it's for local tourists, most of all' and that 'just by being a photo of leisure activity at a luxury hotel, it was already a photograph that defied stereotypes'. DiCampo's main reasoning behind the selection of this image as particularly meaningful was that the photo showed a 'normal' side of life on the continent. He then told us that a couple of years after he took this photo, there was a mass shooting in the

FIGURE 4.2 An image from Everyday Africa portraying children playing in a swimming pool at a luxury hotel in Grand-Bassam, Ivory Coast. Source: Peter DiCampo for Everyday Africa. Reproduced with kind permission.

same spot where 16 people were killed. This tragic event spurred him to repost the photograph 'and say, here's what this exact place looked like on a normal day, so we can remember that the shooting that just took place there is not the norm'. In this regard, DiCampo also added that '[w]e need to treat these things as aberrations, as things that are not normal'.

As a whole, the emphasis on mundanity of Everyday Africa is based on an understanding of photojournalism as capable of repicturing Africa from within through 'evidence of a shared normalcy', as DiCampo stated in his interview with *The New York Times* (Estrin, 2012). In focusing on some of the banal ways in which people's lives are meaningful (see Morris, [1988] 1996), Everyday Africa makes the mundane remarkable, although no single photo in the Instagram feed demands particular attention. With regard to this last point, we now turn to an analysis of plurality as another key storytelling device.

Plurality

When we asked him about whether and how the project breaks photojournalistic conventions, DiCampo highlighted the importance of mundanity. He said that 'the images themselves and what they portray' break with conventions as 'they tend to be quieter, more mundane'. In response to the same question, he also added: 'But the other thing that is quite big is the delivery method'. The fact that Everyday Africa is delivered through Instagram breaks with photojournalistic conventions because we typically tend to associate a particular place or event with a specific image. For example, DiCampo said, 'when you think of the Rwanda genocide, you think of Nachtwey's image of the scarred man'. Instead, Everyday Africa embodies a new form of photojournalism that doesn't rely on the decisive moment or the need to capture the essence of an event through a single image. Rather, it tells a story from multiple perspectives or, as the influential photography expert Stephen Mayes writes in his *Time* magazine article about the project, where 'the truth isn't told in a single photograph' (Mayes, 2017).

Overall, Everyday Africa relies on plurality as a key storytelling device. As DiCampo stated: 'It's ever changing, there are many emotions getting posted to the single Instagram feed within the course of a few days.' To ensure that there is a steady flow to the Instagram feed, DiCampo and Merrill ask contributing photographers to space out their posts by a few hours (Jacobs, 2016). This requirement also contributes to the feed's ability to display multiple perspectives on the continent – for example, where pictures from different countries or focusing on disparate settings and activities are shown side by side (Figure 4.3). Along the same lines, the Instagram feed of Everyday Africa has no central curation. DiCampo explained that he and Merrill wanted to remove themselves from having creative control over the project. He then concluded: '[M]ultiple perspectives are important here, so I'm more interested in being focused on developing the project than having it be my vision, so to speak.'

Communicating visions of collective identity 73

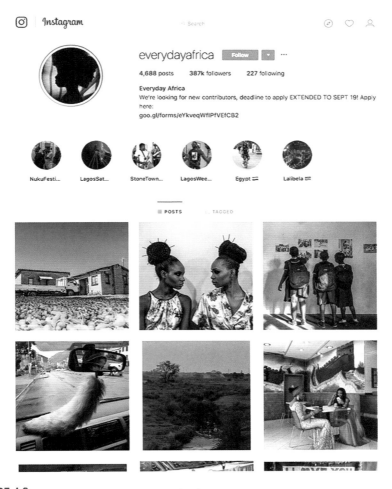

FIGURE 4.3   Everyday Africa's Instagram feed.

In addition, DiCampo told us that for the previous two years or so they had only been adding African photographers to Everyday Africa. The project had developed organically at first, which also made it messy, as 'right away you had more Western photographers than African photographers, you had more photographers clumped in specific countries, for example there are three people in Nigeria and two people in Senegal, and numerous countries with no people'. DiCampo and Merrill felt that it was both important to have a mix of Western and African photographers contributing to the project while also including photographers from as many of Africa's 54 countries as possible. While most African countries had been covered in the project, there were far fewer countries where photographers were based – mainly Kenya, South Africa and Egypt, in addition to Nigeria and Senegal. At the time of our interview, DiCampo had just finalized the project's first call for applications (which was issued in Arabic, French, Portuguese and English) to find additional photographers from across the continent 'that they may have never heard of'. In continuing to develop as a multiperspective narrative from within the continent, Everyday Africa is actively designed to include a plurality of voices, in geographical, temporal and cultural terms alike.

Case study conclusion: seeing Africa as both new and familiar

As a whole, Everyday Africa uses storytelling devices like mundanity and plurality to counter clichés about Africa in ways that promote a more inclusive, yet more intimate approach to representing the continent from an insider's perspective. Because the clichés are rooted in mainstream media portrayals of Africa as a whole, the project purposefully maintains a continent-wide approach to dispelling them (Jacobs, 2016). However, it does so in a way that promotes a highly localized rather than centralized or universalistic approach to representing the continent. Specifically, DiCampo highlighted that 'photojournalism is structured to be about iconic images', whereas Everyday Africa is centred on images that capture otherwise unremarkable moments and small fragments of 'normal' daily life. Furthermore, DiCampo told us that 'the fact that it's a constant stream' also 'serves to take the emphasis off of having anything be iconic and in that way actually in some ways it's more similar to what life is actually like'. Along the same lines, in his *Time* magazine article Stephen Mayes, who has also been a dedicated supporter of the project since its beginnings, wrote that in Everyday Africa, 'what we learn isn't stored in memory in the form of an isolated icon' (Mayes, 2017). Instead, this new form of photojournalism is cumulative, as over time each of these images 'will add a little understanding, contributing to a more fluid and responsive form of knowledge' (Mayes, 2017).

In other words, Everyday Africa is against iconicity. A book that includes a selection of images from the project's Instagram feed was published in 2017, thus pushing Everyday Africa into what could be considered as an attempt to 'fix' some of its images as icons, if it were not for the book's emphasis on the commentary left by the project's followers on Instagram (Acquah et al., 2017). As we saw earlier in this chapter, iconic photography can be a very powerful means for building collective identity, thanks to the emotions that viewers may associate with the recognition of traumatic or triumphal events that have shaped the groups to which they belong. Instead of being centred on the single powerful and often also idealized images that are typical of iconic photography, Everyday Africa privileges the mundane and the plural. Clearly, DiCampo and Merrill's project is a leading example of major transformations in how images are made and distributed, specifically through smartphones and social media platforms like Instagram. Perhaps not surprisingly, then, Everyday Africa has been hailed as a role model for image-makers and educators striving to rise above media-driven stereotypes to create nuanced visual representations of communities and collective identities. For this reason, in 2015 DiCampo and Merrill established The Everyday Projects, a global network of photographers committed to 'creating new generations of storytellers and audiences that recognize the need for multiple perspectives in portraying the cultures that define us' (The Everyday Projects, n.d.).

As a whole, Everyday Africa breaks **photojournalistic conventions** like iconicity and the 'decisive moment', while mobilizing widely shared and familiar visual tropes rooted in humanist photography. Although this project embodies a radically new way of 'doing' photojournalism, we would also argue that Everyday Africa promotes a more traditional humanist ethos, both in relation to the content and the style of its imagery. On one hand, the project's 'everydayness' is a way to emphasize that there are multiple points of commonality among the different places and people that are portrayed as belonging to the same continental community, but also between the portrayed subjects (i.e. 'Africans') and their Western though increasingly also global viewers. On the other hand, these underlying commonalities can be linked to the smartphone aesthetic which 'is casually intimate and connects easily with an audience that may be using the same technology on a daily basis' (Mayes, 2014). Ultimately, as we know from the many times that both DiCampo and Merrill have spoken and written about the project, Everyday Africa is just as much about representing Africa as familiar, as it is about shedding new light on its identity as a continent.

CASE STUDY 2: A RHETORICAL ANALYSIS OF KEY VISUAL STRATEGIES IN THE EUROPEAN CAPITAL OF CULTURE

In 1985, Athens was the first title-holder for the newly launched European City of Culture programme, an idea that had originally been put forward by the Greek Culture Minister, Melina Mercouri. In 1999, this initiative was renamed 'European Capital of Culture' (ECoC from hereon). From 2005 onwards, the ECoC selection process was centralized and each member country was assigned a given year as a host for the year-long capital of culture initiative. Since 2007, two cities have consistently shared the European Capital of Culture title in a move to include the 12 countries that became EU members in 2004 and 2007, together with some countries that were not members of the EU (Table 4.1).

In the new framing of the programme, the European Parliament stressed the need for candidate cities to demonstrate their 'European dimension', which official documents describe as a combination of cultural diversity, shared heritage and cultural cooperation among member states (Commission of the European Communities, 2005). To persuade ECoC judges of their 'Europeanness', candidate cities needed to demonstrate a 'richness of cultural diversity' both in terms of social heterogeneity and the range of cultural activities that they were able to offer (European Parliament and Council of the European Union, 2006). Candidate cities were also actively encouraged to present themselves as being representative of a shared European culture and identity, while being distinctively 'local' and actively engaged in forms of intercultural exchange with communities beyond Europe. Most notably, they were furthermore required to demonstrate 'their links with Europe, their place in it and their sense of belonging' (European Capital of Culture, n.d.: 12). This layered approach to being a capital of culture needed to be effectively communicated in any given candidate city's bid.

To convince panel judges of their 'European dimension' and to keep promoting their status as capitals of culture after a successful application, candidate cities produce a plethora of promotional materials, ranging from official websites and glossy application packets to postcards, posters, bookmarks, pins and even branded clothing lines. The fact that in less than two decades the amount of participating countries almost doubled in number, with most new EU members being countries from the former Eastern Bloc, makes for an even more interesting picture of what a 'European' capital of culture ought to look like.

TABLE 4.1 European Capitals of Culture between 2007 and 2016.

2007	Luxembourg and Sibiu (Romania*)
2008	Liverpool (United Kingdom) and Stavanger (Norway**)
2009	Linz (Austria) and Vilnius (Lithuania)
2010	Essen (Germany), Pécs (Hungary) and Istanbul (Turkey**)
2011	Turku (Finland) and Tallinn (Estonia)
2012	Guimarães (Portugal) and Maribor (Slovenia)
2013	Marseille-Provence (France) and Košice (Slovakia)
2014	Umeå (Sweden) and Riga (Latvia)
2015	Mons (Belgium) and Czech Republic (Plzeň)
2016	San Sebastián (Spain) and Wrocław (Poland)

* At the time of Sibiu's designation, Romania had not yet joined the EU.

** Norway and Turkey are not members of the EU.

Research questions and approach

The two main research questions that we ask here are:

- What are the key visual strategies that were used by aspiring capitals of culture to communicate their identities in promotional materials produced between 2007 and 2016?
- How was 'Europeanness' constructed in these promotional materials?

Here we focus on visual images from the bids and related media produced by 50 candidate cities between 2007 and 2016. This is an especially significant time period, because it marked the inclusion of new EU member states in the initiative, together with the introduction of new guidelines centred on the idea that aspiring capitals of culture ought to effectively communicate their 'European dimension'. We collected materials from aspiring capitals of culture over several years, creating an archive of both digital and analogue media as each city's ECoC candidacy unfolded. Using a rhetorical approach, we grouped the main **visual strategies** that we encountered across promotional materials from our archive to then discuss some of their implications for an understanding of these cities' identities in the face of EU enlargement and globalization.

Findings: multiplicity, metonymy and abstraction for three key identity claims

Given the selection process that underlies this EU-led initiative as well as its implications for a city's visibility within Europe, together with local urban regeneration and the influx of global tourism, in these promotional materials aspiring European capitals of culture perform their identities through multiple rhetorical strategies. Specifically, here we examine multiplicity, metonymy and abstraction as the main visual strategies used by cities to perform intersecting identity claims, including local, pan-European and globalizing dimensions.

Multiplicity as diversity: communicating pan-European identity

First, we identified multiplicity as a visual strategy aimed at communicating cities' identities as diverse, in order to both fulfil and further reinforce the dominant political discourse on European identity as being grounded in diversity. This is a visual strategy that exploits what Tufte (1997) calls the 'smallest effective difference' – that is, minimal variations in visual content to achieve maximum effects of differentiation. In other words, multiplicity maximizes on the 'appearance' of richness, variety and thus also diversity, regardless of how truly varied the representational elements used in an image's layout happen to be (Aiello, 2007).

There were two main kinds of multiplicity that could be found across candidate cities' promotional materials – namely, multiplicity of colours and multiplicity of images. Multicolour schemes were predominant, particularly in the design of logos. The logo for Lübeck 2010 used a repetitive pattern of the same geometric shape – a square – which was rendered in different colours. Likewise, the logo for Segovia 2016 was a multicolour abstract motif. With a slightly different approach, a Flash intro in the official website for Marseille-Provence 2013 communicated the candidate city's diversity multimodally through an animation of staggered and variably long and thick multicoloured bars which appeared from left to right undulating in a wave-like motion (see a screenshot in Figure 4.4, together with Lübeck 2010 and Segovia 2016's logos). In a related manner, the first website for Istanbul as a 2010 capital of culture included a Flash animation in which the logo changed colour in a loop.

FIGURE 4.4   Multicolour schemes in the logo for Segovia 2016, the official website for Marseille-Provence 2013, and the logo for Lübeck 2010.

Visual layouts featuring multiple images were also widely used. This approach was most often found in print materials that called for a high concentration of information in a limited amount of space. A number of postcards, posters and even bookmarks were set apart by 'busy' collages of images (Figure 4.5). It is important to point out that this multiplicity of images did not correspond to a high degree of variety or diversity in their representational content. For example, the inside cover of Turku 2011's bid book featured a visual layout that was made of multiple images of people who were almost exclusively white, and more specifically Nordic, with the exception of only two portraits of children who were distinctly non-white and ethnically non-European. In a similar fashion, Halle 2010's postcard was set apart by the uniformity of its contents, which over-represented whiteness and masculinity together with a particular vision of culture as 'high culture' (Aiello and Thurlow, 2006).

Often, the two types of multiplicity that we have just described worked off each other to amplify this 'perception' of diversity. These promotional materials frequently included elaborate compositions featuring combinations of multiple colours and images. In addition, these visual compositions were rhetorically reinforced by linguistic claims related to a city's cultural diversity. In its bid to

FIGURE 4.5 Multiplicity of images in a postcard layout for Halle 2010 and the inside cover of the bid book for Turku 2011.

become a capital of culture in 2011, Rovaniemi described itself as a place 'rich in forms, languages, layers and faces' (Rovaniemi 2011, n.d., para. 1). Likewise, the bid for Potsdam 2010 described the city as 'a melting-pot of European ideas and thoughts' (Potsdam 2010, Concept section, para. 2).

Metonymy: framing the local as European

In addition to highlighting these cities' diversity, ECoC promotional materials made claims about their European identity through metonymy, a visual strategy that contributes to framing candidate cities' local identities as more broadly representative of their 'Europeanness'. Metonymy is achieved by evoking a totality through one of its often physical or material parts, and thus also entails an act of symbolic reduction (Burke, 1952). In classical rhetoric, for example, the Greek god of wine and festivity, Dionysus, is often represented by his main instrument, namely a cup (Eco, 1984). Overall, metonymy is defined by a relationship of interdependence between a given totality and the rhetorical markers that are taken to stand for it as a whole. In visual communication, metonymy often results in the adoption and repetition of 'key icons' (Scarles, 2004) connoting a given identity, as this relates to a culture or a place. Due to the narrow selection and symbolic reduction that are inherent in their production, visual metonymies often result in highly controlled and sanitized images.

The promotional materials produced by aspiring capitals of culture abound in images that selectively show parts of these cities, and specifically parts of these cities' buildings that are commonly considered typically, if not prototypically, European, such as columns, Greco-Roman capitals and reliefs, arches and domes. 'Snippets' of buildings highlighting some of these very details were featured in several website banners (Figure 4.6). Likewise, bid books often featured selected images of 'old-world' cityscapes and vistas such as cobblestone pavements and nineteenth-century building façades. The bid book for Pécs 2010 – which was designed by nationally renowned Hungarian graphic designer Zsolt Czakó – contained several pages featuring abstract designs of geometrical shapes such as circles and rectangles, with each of these being a 'window' into a metonymic photograph of an architectural detail. By cropping out details that may have been associated with some of the more specific characteristics of a city's local context, these metonymic images framed aspiring capitals of culture as 'generically' European.

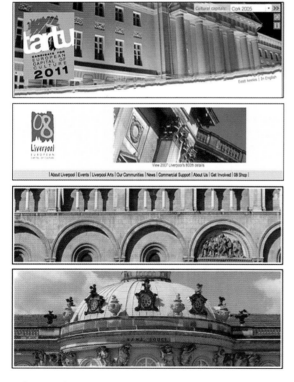

FIGURE 4.6 Metonymic architectural details in banners for the official websites of Tartu 2011, Liverpool 2008, Pécs 2010 and Potsdam 2010.

Abstraction: fashioning the local as global

Finally, we found that cities used visual abstraction to represent themselves as 'global' rather than just European. Abstraction is a visual strategy that relies on what the famous art theorist Rudolf Arnheim defined as the 'priority of simple, global forms' (Arnheim, 1947: 69). Abstract motifs are often associated with symbolic or conventional meanings. However, perceptual understanding is also made easier by the presence of simple patterns; for this reason, abstract painting has been used as a method to portray universal, absolute truth (Zimmer, 2003; Cheetham, 1991).

Across aspiring capitals of culture, promotional imagery and logos in particular use abstraction to perform an identity that reaches beyond a city's Europeanness, and to fashion its local specificity as global. For example, the logos for Liverpool

2008, Istanbul 2010, and Essen and the Ruhr Region 2010 were all stylized renditions of distinctively local visual content (Figure 4.7). While Essen and the Ruhr Region's logo was centred on an abstract image of the geographical area covered by the Ruhr in Germany, Liverpool's logo was an impressionistic representation of the city's world heritage waterfront. An official press release for Liverpool 2008 stated that the logo 'appeals to a global audience and is immediately recognisable as Liverpool, European Capital of Culture, 2008' (New 08 brand, 2004). In the same press release, Liverpool 2008's marketing director likened the logo to 'other internationally recognised brands for major events such as the Olympics' and stated that it would help the city 'attract investment and tourism'. We also spoke with the designer behind the logo for Istanbul 2010, who highlighted this visual concept's ability to combine locally distinctive traits with more modern and global qualities. The logo was an abstract rendition of Istanbul's hills and bridges, but the designer emphasized that its roundedness also stood for the 'curvy' nature of Turkish culture. In sum, these logos used abstraction to link a candidate city's specific, local identity to the streamlined and stylized aesthetic that is typically associated with global brands.

FIGURE 4.7   Stylization as abstraction in the logos for Liverpool 2008, Istanbul 2010, and Essen and the Ruhr Region 2010.

Abstraction was also often achieved through the exclusive use of typography in logo design (Figure 4.8). Linz 2009's logo was chosen by a selection jury because, as a visual rendition of the numbers 0 and 9 through oversized punctuation marks, it was considered to be versatile enough to 'remain as fresh and interesting' over time and across contexts (Der Weg zum Logo, 2006). Official press releases emphasized the logo's ability to fit into a variety of visual scenarios, particularly photographs portraying some of the artistic, industrial and natural resources of Linz, where a variety of round objects were made to replace the logo's point mark next to the comma mark. Along the same lines, the logo for Plzeň 2015 used a five-dot motif that was later juxtaposed to a variety of images of local sights and events. Pécs 2010's logo is another prime example of how abstraction was used as a key visual strategy. The city's bid book detailed that the logo design was purposefully based 'on the modern, constructivist tradition of art in Pécs and refers to the architectural practice of the city's Bauhaus school' while also being rooted in the latest aesthetic developments in contemporary design – that is, 'a more rational, simpler visual language more closely related to modernism' (*Borderless City*, n.d.: 122, emphasis added). As a whole, in ECoC promotional communication abstraction was used to fashion a city's specific identity as globally appealing.

FIGURE 4.8 Typography and abstraction in the design of logos for Linz 2009, Plzeň 2015 and Pécs 2010.

Case study conclusion: being local, European and global

The three key visual strategies that we have just examined point to the multilayered identity claims that candidate cities make in an attempt to persuade both panel judges and the public of their worthiness as potential capitals of culture. Our archive included promotional materials from cities with disparate economic resources, historical backgrounds and cultural cachets. Yet, they all equally and independently subscribed to a shared visual communication strategy in order to look like 'proper' European capitals. There is clearly a dialectic at work in these promotional materials, where cities are keen to showcase their specificity while also fitting into an overarching aesthetic for symbolic and economic gain. This rhetorical work entails not only the selective inclusion of locally available representational traits, but also their stylization towards a generically European and globalizing 'look'. In doing so, ECoC promotional materials do not subscribe to a strictly homogenized version of 'Europeanness'; rather, they exploit the local specificity and distinctiveness of these cities to fashion them as representatively European and even world-class through a set of common rhetorical strategies.

What emerges here is a Europe whose unity is most often communicated via metonymic references, or through a selection of specific places that are then visually framed so as to appear as 'prototypically' European. By invoking classical, high culture as European, the visual images used in these promotional materials contribute to particularistic and therefore also exclusionary narratives of European identity (cf. Aiello and Thurlow, 2006). This is also a Europe where diversity is key to a shared sense of belonging, but where concrete examples of diversity are actually hardly ever shown and, when they are made visible, they are stylized to fit a globalizing aesthetic or are altogether exoticized, thus resulting in strategic self-othering as in the case of Istanbul, for example. Ultimately, the vision and version of Europeanness that is communicated in these promotional materials is riddled with tensions between differences that may be included and even actively exploited in the pursuit of symbolic and economic gain, and those that may instead be less becoming of a quintessentially European capital of culture and therefore ought to be left out or communicated as 'other' (see Hall, 2003).

CHAPTER CONCLUSION: VISUALIZING A SENSE OF BELONGING BEYOND NATIONAL IDENTITIES

In this chapter, we have examined some of the ways in which imagined communities like 'Africa' and 'Europe' are communicated visually. In particular, our case studies have focused on key storytelling devices used in the Instagram-based project Everyday Africa and on the rhetorical strategies used by aspiring European capitals of culture in their visual promotional media. Across the two case studies we have noted that image-makers agentfully and strategically use particular visual techniques to **perform identities** in ways that respond to the assumptions or needs of both actual and implied audiences. The outcomes of these image-making practices are, of course, always ideological, while also being tied to the communication technologies and genres that underlie the production and distribution of particular types of images (e.g. Instagram photography or logo design). Overall, this chapter contributes a nuanced understanding of the relationship between visual communication and both the formation and maintenance of collective identities. By focusing on the storytelling devices and rhetorical strategies that are key to promoting particular identity claims, we can begin to think of the multiple ways in which the same imagined community may be constructed according to different goals and perspectives.

POTENTIAL FURTHER RESEARCH

The approaches that we discussed in this chapter could be extended to a variety of media and types of imagined community. As we mentioned earlier, collective identities come in many shapes and forms beyond the construct of the nation. It would therefore be productive to investigate some of the ways in which particular ethnic groups within

(Continued)

a country (like, for example, Native Americans in the United States, the Māori in New Zealand or the Uyghurs in China), or diasporas within and across different countries (like diasporic South Asians or Chinese in Britain, South Africa or the Philippines) represent themselves visually on the internet and/or in community media (e.g. newspapers, film and television).

Another way to explore the role of visual images in the construction of collective identities would be to focus on how members of a particular national or ethnic group relate to visual portrayals of their community. Key questions here could be: 'How do members of the same group describe and interpret the images found in visual media that aim to represent their shared collective identity?' and 'Are the ways in which the media portray a particular group central to how its members think of themselves and of each other as belonging to the same imagined community?'

FURTHER READING

Important sources on the relationship between iconic or humanist photography and collective identities include Hariman and Lucaites (2007a), Hamilton (1997), and Lutz and Collins (1993). For research on the role of visual communication in the promotion of European identity within the EU, see Fornäs (2012) and Aiello (2007 and 2012a).

5 WAYS OF SEEING DIFFERENCE BEYOND STEREOTYPES

This chapter considers scholarship on difference, othering and stereotypes in the media, with an emphasis on how the *pictorial* aspects of representation hold particular power to injure (Lester and Ross, 2003), even if not deliberately intended. We summarize visual media's role in enacting categories of difference and why this representational practice is deemed harmful. But we also examine how varied media forms can play a leading role in challenging stereotypical and **reductive portrayals**, with the potential effect of shifting societal attitudes towards marginalized communities.

In sum, the chapter:

- discusses how images enact social and cultural difference;
- explains how theoretical contributions from cultural studies and postcolonial theory inform visual media analysis;
- examines how pictorial stereotypes are shaped by power relations and societal inequalities, and how representational shifts can in turn shape social attitudes;
- conducts an analysis of the title sequence for Netflix drama *Orange is the New Black* to examine how it visually celebrates difference and diversity in its self-branding (Case study 1);
- analyses the visual campaign strategies of Save the Children to explore how charities attempt to harness humanizing imagery without 'othering' the children they aim to help (Case study 2).

HOW MEDIATED IMAGES ENACT SOCIAL DIFFERENCE

Media forms can engender a sense of belonging – for example, in Benedict Anderson's (1983) writing on 'imagined communities', which referred to the way in which national daily newspapers construct a nation and sustain nationalism through addressing 'an imagined political community', or, more recently, in the online communities where support groups can flourish despite geographical distance (see Chapter 4). But in creating a recognizable 'us' there is often an implied 'them': the 'out-group' that might simply be absent from media representations, or marginalized, or directly portrayed in a negative light. Visual images are particularly adept at enacting social difference, at reducing people to types, or to symbols or metaphors for wider social issues. Likewise, as we've already explored in the preceding chapters, identity formation and performance could be said to be increasingly dependent on mediated images (Sturken and Cartwright, 2017; Hall, 1997; Woodward, 1997).

When we write of 'ways of seeing' difference, we are indebted to the superb writing of art critic John Berger. Berger's 1972 television series and accompanying book, *Ways of Seeing*, turned traditional art history on its head and urged viewers and readers to question the common-sense assumptions about status and desirability perpetuated in both art and advertising. Influenced by critical theory, semiotics, cultural studies and postcolonial theory, Berger's concern is how systems of power operate in the cultural sphere, specifically here in the realm of visual culture. As Berger (1972: 129) explored, our 'ways of seeing' are both socially situated and individual, as we respond to a 'density of visual messages', many of which impose narrow ideas of desirability and glamour.

In his discussion of works of art, Berger demonstrates how both the artist's choices and the way people look at the artwork are affected by assumptions relating to 'beauty, truth, genius, civilization, form, status, taste, etc.' (1972: 11). By examining how this 'way of seeing' is embodied in the image, we can start to interrogate the sometimes disturbing presumptions revealed – how status and power are reproduced and bolstered, how the spoils of colonial pasts are represented, how the poor and people from non-Western countries are depicted in relation to their white counterparts, how women become objects surveyed by men. Along with other scholars writing in the same era, Berger helps us to understand how images play a significant role in constituting the world around us, and that a critical eye can reveal unsettling traces of colonial-era prejudices, reductive perspectives of other cultures and gendered depictions that justify patriarchal systems of control.

In the next section we explore how, in recent years, media representations of Islam and Muslims in particular have demonstrated the ways in which marginalization and negativization of certain communities remain a disturbingly stubborn characteristic across Western cultures.

Defining the other: Orientalism, Islam and race

Articles on 'othering' in news media often focus on the linguistic elements and the discursive strategies employed (whether knowingly or unthinkingly). In recent years, for example, Muslim people living in the West have faced regular negative misrepresentations in the press and other media. As charted by others (Poole, 2002; Alsultany, 2012), this form of prejudice has a longer history, but the 9/11 attacks and the US administration's 'War on Terror' response to those attacks strengthened the fear-inducing 'us' and 'them' rhetoric, especially towards those from Middle Eastern countries (broadly and often erroneously defined). While Islam is a religion shared by people of many ethnicities and nationalities, the negative portrayals are often intricately entwined with racist assumptions associated with the idea of 'the Orient'.

Much of the work in this area is influenced by the writing of Palestinian-American scholar Edward Said. One of the godfathers of postcolonial theory, Said's book on *Orientalism* (1978) presented a critique of the academic field of Oriental Studies, pointing out how, even when sympathetic in its tone, it was based on the premise that European and US scholars knew more about the Orient than those who lived there – in other words, authoritative knowledge of the Islamic world was perceived through a colonial and imperialist lens. This **system of difference** identified by Said is premised on a clear distinction between the West (the 'Occident') and the East (the 'Orient'), with those from the West misrepresenting the East as an alien culture through essentialized caricature. Crucially, casting Islam as a threat to the West or (paradoxically) as weak and backward, works to justify military aggression or other interventionist policies.

Said commented more directly on contemporary media representations in his book *Covering Islam* (Said, [1981] 1997) and a number of studies have contributed to our understanding of how both factual and fictional representations reproduce lazy stereotypes. For instance, Elizabeth Poole's book *Reporting Islam* (Poole, 2002) looks at how anti-Muslim sentiment is expressed in UK tabloid and broadsheet newspapers alongside audience responses, while Jack Shaheen's (2001) study of Hollywood films found over 1,000 films made between 1914 and 2001 in which Arabs are stereotyped as evil-doers and villains.

We would argue that the emotive power, symbolism and 'racialized aesthetics' (Sturken and Cartwright, 2017) deployed in photojournalism, among other visual media genres, deliver amplification of difference and danger in a worryingly effective manner. In the years since 9/11 the conflation of 'Islam' with 'terrorism' in the news media and beyond homogenizes the Muslim identity, fuels hate crimes, and undermines efforts to tackle alienation and extremism (Saeed, 2007). Islamophobic portrayals that cast Muslims as 'other' and 'foreign' serve to polarize communities and provide succour to far-right groups whose open hostility and violent response appears to be mounting, whether in the UK, US, Hungary, Germany or Australia (Akbarzadeh, 2016).

It is not only in traditional mass media where we observe patterns of identity formation and othering. More recently, researchers have turned their attention to discursive patterns in social media messages – for example, examining the way in which certain identities and group memberships are constructed on Twitter through the use of hashtags. In an instance of positive identity formation, Sanjay Sharma (2012) studied how 'blacktags' constitute Black Twitter identities, while an exclusionary example can be found in Ramona Kreis's (2017) study of racist online discourse regarding immigrants and refugees that coalesced around the #refugeesnotwelcome hashtag in the context of the European refugee 'crisis'. Importantly, Kreis conducts a multimodal discourse analysis, making the point that visuals can support the naturalization of racist discourses with visually striking or emotive images: 'It is thus easy to deny visually communicated xenophobia; the invoked message can be dismissed since it is not explicitly present' (Kreis, 2017: 6). In such pictures, refugees are found by Kreis to be depicted repeatedly as violent criminals.

Studies of this nature are in dialogue with Stuart Hall's earlier writing on popular representations of racial difference, where he explains how naturalization is a representational strategy designed to '*fix*' difference 'and thus *secure it forever*' (Hall, 1997: 245, emphasis in original). We draw upon this work in the next section on stereotyping.

'Fixing' difference: stereotyping, marginalization and inequality

The idea of 'fixing difference' is most blatantly observed in the representational practice of stereotyping. Stuart Hall's chapter on 'the spectacle of the other' in *Representation* (1997) provides an excellent introduction to how stereotypes work. While recognizing that it is difficult to make sense of the world without employing categories and types to distinguish objects or to describe people, this becomes problematic when there are inequalities of power and even 'symbolic violence' exercised towards a group of people: 'Stereotyping reduces people to a few, simple, essential characteristics, which are represented as fixed by Nature' (Hall, 1997: 257). The rigid nature of stereotypes is designed to exclude 'others', to highlight and separate the deviant ('them') from the normal ('us'), to impose essentialized and oversimplified characteristics that usually come with connotations of inferiority. The 'fixing' of difference comes with a clear sense of hierarchy, where those with the power and capability to represent are self-appointed as embodying the superior values, morals and societal progress.

However, Hall recognizes that this is not the entire story. Stereotypes, like all forms of representation, are never entirely fixed in meaning. We notice stereotypical representations because they have a certain 'stickiness' and endure, but once familiar, they also become contested and challenged. One counter-strategy is to eclipse the prevailing negative stereotypes with more positive and diverse representations. Campaigns with the specific intent to challenge stereotypes often play with viewer expectations – the apparent teenage mugger who is actually chasing the robber, the ace footballer who is a woman, the graceful ballet dancer who has a prosthetic leg – there are countless examples that aim to challenge audience presumptions. But there are limitations to a gimmicky 'reveal' strategy, and without addressing the inequalities of power relations that underlie stereotyping practices, meaningful representational shifts take time and require the embrace of complexity and diversity both on screen and in cultural production.

Visibility is a key concern here. Gaining **visibility** in the media can be viewed as redressing an imbalance for those previously marginalized, as remedying an injustice, and, for those able to take control of the nature of their visibility, as gaining legitimacy as recognized social actors. For example, it is only in recent years that people identifying as trans have gained growing and diverse representations across popular media (see Chapter 3). The increasing range of dramatic representations have led to public debates about whether only transgender actors should be employed in such film and TV roles. In a sign of shifting notions of acceptable casting, in 2018 Scarlett Johansson backed out of playing real-life transgender character Dante Tex Gill, who identified as a man, in the film *Rug and Tug* (2019), following backlash from the trans community. Johansson had already faced criticism in 2017 for her casting in the adaptation of Japanese anime classic *Ghost in the Shell* (2017), with the production accused of 'whitewashing' instead of casting Japanese actors.

These two brief examples point to the continuing battles for *visibility* and *legitimacy* fought by underrepresented groups, and why the nature and prominence of representation for such communities remains culturally significant and contentious. Popular media representations play a role in how we see ourselves, and how we place ourselves within social spaces and political structures. To be excluded or portrayed in a distorted manner can have real-life consequences. The examples also highlight contemporary debates about who is actually employed in the cultural industries and their capacity to influence the kinds of cultural artefacts produced. Increasingly, appeals for diversity are not restricted to on-screen portrayals and employment for actors from marginalized groups; they also interrogate the diversity of writers, directors and production crew – for example, see Anamik Saha's book (2017) on race and the cultural industries.

IMAGES THAT INJURE AND MOVING BEYOND

In their edited collection, *Images that Injure* (2003), Paul Martin Lester and Susan Dente Ross bring together a number of case studies about stereotypes, categorized as '9/11' (referring to the terrorist attacks on 11 September 2001), ethnic, gender, age, physical, sexual orientation, and also a miscellaneous section including politicians in cartoons, geeks and media personnel (this list refers to the second edition). Their overall focus is on the way in which media images interact with the mental images we have of certain groups of people. In their introduction, the editors make the case that '[p]ictures are highly emotional objects that have long-lasting staying power within the deepest regions of our brain' (2003: 3). The book offers an excellent introduction to pictorial stereotypes and the ethical concerns that underpin scholarly investigations. The collection mostly covers stereotypes in US culture and interestingly there is little attention to social class – something that receives much more consideration in the UK context.

Below, we summarize how some of the key categories of pictorial stereotypes are researched in contemporary media forms (initially using those categories in Lester and Ross's collection). Crucially, such studies not only signal the importance of understanding how meaning is constructed in specific media genres (such as comedy), but are also sensitive to how such representational practices are connected to the political economy of the media industries and wider societal inequalities.

- *Ethnic minorities and race* This is probably one of the most studied forms of stereotyping or 'othering', often considered alongside other factors such as markers of poverty or religion. As already discussed in this chapter, discourses of race often work to categorize and divide people: 'they signify in the way they mark out human differences' (Hall, 2017: 31). In their study of images in news framing, Paul Messaris and Linus Abraham (2001) argued that visuals in particular can deliver subtle racist cues that evoke well-worn stereotypes of African Americans. Video games have long received academic attention for their problematic depictions of race and gender – for example, in the reification and 'flattening out' of Arab and Muslim identities often appearing as the 'enemy'. However, Vít Šisler (2008) and others have started to explore how games made in the Middle East are challenging misrepresentations and offering more culturally balanced player experiences.
- *Gender* This is another area that is well covered, with especially female representations in media imagery examined for characteristics of gender stereotyping and objectification. For instance, the roles performed (wife, mother, sexual creature), 'feminine' gestures and bodily positioning, as famously explored in Erving Goffman's (1976) study of advertising, in addition to the imagined female audience as an avid consumer of 'beauty' products. Studies of media constructions of masculinity are now also common, covering everything from the dramatic portrayal of heroes and anti-heroes (Bruun Vaage, 2015), to anxieties about the relationship between toxic masculinity and violence.
- *Age* Older people have traditionally been subject to negative media portrayals, often manifested in relation to poor health or narrow ideas of age-appropriate behaviour. Indeed, older women can feel invisible both in daily interactions but also in the media. As economic circumstances shift and people live longer, people of retirement age and above are becoming recognized as a key target audience both for advertisers and programme-makers. Television series such as BBC1's *Last Tango in Halifax* (2012–16) have attracted praise for showing people in their seventies in romantic relationships, and with complex and rich characterizations. The programme also received acclaim for its representation of class and LGBTQ (lesbian, gay, bisexual, transgender, queer) relationships.
- *Physical* A growing research interest in disability and the intersection with media representations can be observed in journals such as *Disability and Society*. One recent publication, a book by Alison Wilde (2018), examines the construction of cultural representations of disability in comedy and how this contributes to shaping both wider cultural attitudes and disabled people's

images of themselves. Comedy is one of the genres perhaps most susceptible to employing stereotypes to gain easy laughs, but comedy can also take risks in satirizing stereotypes or using irony as critique – examples provide rich and often contradictory texts for analysis. The line between 'laughing at' and 'laughing with' is a tricky one to determine, and audience research offers a valuable way to explore how those represented engage with such portrayals. The term 'physical' here could also encompass the way in which learning disabilities, mental health conditions or psychological trauma are represented in documentaries and reality TV, or performed by actors in dramas.

- *Sexual orientation and gender identity* As observed above, the diversity and positivity of roles for LGBTQ protagonists have started to challenge the predominant heteronormative paradigm in media images, reflecting societal changes such as the legalization of same-sex marriage. That is not to deny the existence of continued prejudices and occurrences of homophobic and transphobic violence in many countries where attitudes have not necessarily caught up with legality. Reception studies examining how groups respond to LGBTQ stereotypes can be commonly found in journals such as *Sexualities* and the *Journal of Homosexuality*, together with longitudinal content analysis that, for instance, shows how increasing visibility in advertising is accompanied with a conformist domesticized version of 'gayness' (Nölke, 2018). Increased visibility for transgender people in the media has arguably improved societal recognition of different gender identities, in addition to sexual orientation, and we are hopeful that respectful portrayals play a part in furthering acceptance and understanding.
- *Social class* In their edited collection, *Social Class and Television Drama in Contemporary Britain* (2017), David Forrest and Beth Johnson posit that television remains a significant site for the formulation of class identity and emphasize that 'television texts can be understood not as static, but as operatives or envoys of change' (2017: 4). In this spirit, a number of the essays explore how drama series such as *Peaky Blinders* (BBC, 2013–) have reintroduced dimensions of class, gender and regional identity to the national conversation. Notions of taste and aesthetic beauty are also intricately bound up with class status, with the (upper) middle-class perspective often naturalized as authoritative and desirable.

The above summaries also hint at how these categories intersect and are interwoven in media images. In addition to the above 'master' categories, paying careful attention to other more specific **binaries** and stereotypes provides clues into the biases and prejudices that are sometimes unconsciously replicated in media portrayals. These take different shapes across varying cultural contexts, but we can also observe how they potentially travel across global media culture. Some possible examples include: rural versus urban; religious identity; education levels (including 'geeks'); certain professions or jobs; mental health; body shape; Millennials or 'digital natives'; models of parenting; refugees and immigrants, etc. What other categories of noteworthy pictorial stereotypes can you think of?

The next two case studies examine contemporary examples encompassing very different media forms – a television drama title sequence and a charity appeal – to highlight the ways in which producers of such content attempt to move beyond the stereotypical depictions associated with each genre.

CASE STUDY 1: HOW *ORANGE IS THE NEW BLACK* PROMOTES DIVERSITY AND DIFFERENCE IN ITS TITLE SEQUENCE

Orange is the New Black (hereafter *OITNB*) premiered on Netflix in July 2013 and at the time of our research its sixth season had been released (in July 2018). The drama is based on the prison memoir by Piper Kerman, fictionalized in the Netflix series as Piper Chapman. Sentenced to 15 months in a low-security penitentiary for transporting drug money for her girlfriend ten years earlier, Piper's prison experience allows the creators to tell the stories of incarcerated women in the United States through the eyes of an upper middle-class privileged blonde woman.

As recounted in an interview and much cited since, series creator Jenji Kohan knowingly used the relatability of the 'girl next door' to explore diverse women's stories while recognizing Piper as a 'useful' way to sell the show to the network.

> In a lot of ways Piper was my Trojan Horse. You're not going to go into a network and sell a show on really fascinating tales of black women, and Latina women, and old women and criminals. But if you take this white girl, this sort of fish out of water, and you follow her in, you can then expand your world and tell all those other stories. (Kohan interview cited in Farr, 2016: 168; see also San Filippo, 2017: 76)

Kohan is clearly disparaging of the networks for not being open to stories of black/Latina/old/criminal women, and yet the success of the series suggests a huge audience for a diverse female-led cast. Netflix do not release viewing figures for individual shows, but the series is both critically acclaimed (Emmy-winning) and considered one of the most successful and talked about of the Netflix Originals series. Maria San Filippo (2017: 80) comments on how Netflix's strategy of releasing whole series at one time resembles the temporality of the blockbuster opening weekend, rather than the scheduled pace of a series, with online viewer communities displaying 'highly affective immersion and devotion, both in spectatorial practices of gazing, bingeing' and in fan-interactions on social media.

Setting a drama in prison already offers the writers a promising vantage point through which to explore societal issues and institutional power structures. As a microcosm of how a society deals with 'social ills' (crime, racism, drug addiction, poverty), prison dramas are able to problematize normative notions of moral and legal fairmindedness in the criminal justice system through humanizing those incarcerated and revealing harsh inequities. As others have observed, including author Piper Kerman who now campaigns for justice reform, the conditions for women in prison and the experiences in life that led them there, are growing concerns in the context of increased imprisonment of women in the United States (http://piperkerman.com/justice-reform/; see also Farr, 2016). Indeed, the US has the highest rates for female prison population in the world.

Piper Kerman and Jenji Kohan's public statements give us insights into their motivations for representing a diversity of women's lives in prison. It is worth noting that this is foremost an entertaining comedy-drama rather than a gritty realist exposé, but their comments suggest the series creators aim to take representational risks by telling stories beyond Kohan's blonde white 'Trojan Horse'.

Research questions and approach

In this case study we focus on the opening title sequence for *OITNB* as a window into the distilled essence of the programme. As defined by Monika Bednarek (2014: 126), the television title sequence 'is a relatively short, recurring multimodal composite, usually including music and the title of the television series and often also credits'. Bednarek notes that television title sequences have not received a great deal of scholarly attention, but they are 'key cultural products that viewers clearly do engage with and enjoy', as demonstrated in the online galleries and websites dedicated to the 'coolest' and 'most inspired' (2014: 127). Bednarek relates how television producers take pride in creating visually appealing title sequences that skilfully capture the ethos of the programme, benefiting from graphic design developments such as Photoshop and animation tools.

Whereas television title sequences have so far been underexplored, the aesthetics, functions and conventions of cinematic film title sequences have received attention as a unique 'crossover' space, 'from the outer to the inner side of fiction' (Stanitzek, 2009: 57). Georg Stanitzek proposes the film-title sequence as a self-reflexive reading of the film itself, set apart in its miniature and inventive form: 'insofar as it is endowed with its own beginning and end, it establishes itself as distinct and develops its own coherence' (2009: 45).

In Netflix's binge-watching oriented design, where the consumer chooses the pace of viewing, title sequences can be 'skipped' similarly to a YouTube advert, but we argue that such sequences present a brand identity and condense the programme's quintessence in potentially revealing ways. Our primary guiding question for this case study is:

- What are the visual discursive strategies used in the title sequence for *OITNB* and how do they work to convey the essence of the series?

To break this down further into questions that could be applied in any title sequence analysis:

- In what ways does the title sequence perform its conventional functions: crediting authorship, producers or actors; introducing characters or narrative elements; establishing mood; creating brand recognition?
- How does it innovate or surprise?
- In 'reading' the title sequence as a self-reflexive rendering of the programme (Stanitzek, 2009), does it effectively capture its essence or ethos?

Our interpretation is informed by background reading into *OITNB*'s critical and commercial reception, and by viewing the series itself with a **thematic focus** on issues of diversity and difference. The above questions are designed to interrogate the ways in which a specific visual discourse is constructed in the title sequence and the techniques employed to exemplify the ideological values of the programme. Following Fairclough (1995a), we investigate the 'traces' of discourse practice in the branding of *OITNB* by describing and analysing how the title sequence conforms to certain intertextual expectations, and also subverts them.

Findings: celebrating female diversity in the *Orange is the New Black* title sequence

The *OITNB* title sequence lasts 71 seconds, which is relatively lengthy in comparison to those included in Bednarek's (2014) quantitative survey of 50 fictional series. In further research, it would be interesting to see how this compares to other Netflix originals or more recent quality drama series. In terms of conventional features, the ownership is first signalled by 'Netflix presents' appearing on a black screen. The theme tune by Regina Spektor ('You've Got Time') then kicks in. The next written titles to appear are 'starring taylor schilling' (no capitals), with other actors' names appearing at regular intervals (laura prepon, michael j. harney, michelle hurst, with kate mulgrew and jason biggs). Finally, the name of the programme appears: 'ORANGE is the NEW BLACK' (in logo style), 'created by Jenji Kohan'. Only a few named stars and the creator are therefore credited within the sequence, fulfilling the 'naming/identifying function' (Bednarek, 2014) but in a minimal style.

The music is a central feature, as is often the case in title sequences, which rarely include spoken word or dialogue. In the case of *OITNB*, the music was composed especially for the series, opening with a guitar sound and a forceful, almost violent style of singing. The lyrics are inspired by rough cuts of the series' content, and speak directly to the fears and frustrations of a prison environment (Regina Spektor interviewed in Pirnia, 2013). The performer Regina Spektor is an acclaimed singer-songwriter with a distinctive, indie, anti-folk musical style whose work has appeared in various media and advertising. Spektor signifies a combined coolness and TV-friendliness. The only other sound is the locking of heavy doors, working to place the viewer figuratively inside the prison.

However, we want to focus attention on the more surprising visual characteristic of the title sequence. Eschewing any kind of narrative sequence or clips from the series action itself, the title sequence offers a series of (moving) portraits, all shot in the same extreme close-up, revealing either the eyes or the lips. This sequence is only broken by the following snapshots: female hands in handcuffs; fingerprinting; an orange jumpsuit-clothed torso with a strong light centred on her chest; a wire fence with a visiting hours sign; three phones on a wall; a watchtower; and finally, the top of a prison wall with razor wire as the programme name appears. These are clear allusions to prison life but they remain at an abstract level. Other than those few stock prison images with no or minimal movement, the extreme close-up portraits dominate. But these are not necessarily portraits of the cast. If the cast do appear, it is so fleetingly that they are indistinguishable from the other faces. Most of the faces we see are real-life former prisoners.

The sequence was designed by the Thomas Cobb Group (TCG), who had been instructed that Kohan wanted the credits to tell the stories of multiple women's lives, not just Piper's tale. Over 50 women were photographed, 'found via Homeboy Industries, an organization that helps the previously incarcerated and gang-involved redirect their lives with education and employment services, therapy, tattoo removal, and case management' (Dunne, 2013). The original memoirist Piper Kerman also appears as one of those recorded (she blinks and therefore is one of the 'moving' portraits). The extreme close-ups manage to protect the identity of those photographed, while presenting an intimate and non-airbrushed depiction of a diverse group of women. In the next section we draw upon the literature on photographic portraiture to explore how we might interpret this visual discourse approach.

Portraiture as the dominant visual strategy in the title sequence

Title sequences are generally considered to settle the viewer into the imaginative space of the show's creators. The decision to use 'real women' rather than the fictional characters is therefore deemed a 'radical' departure (Dunne, 2013). The sequence design positions the viewer as looking directly into the faces of a diverse range of women of different ethnicities and ages – their pores, freckles, piercings, tattoos and wrinkles visible (see Figure 5.1). The strong central lighting highlights each facial feature, with a black background and shadow surrounding the face. The rapid editing means that we rarely linger on any single face, although still photographs are mingled with moving images. As the faces smile, laugh and blink, the 'flat' quality that is often associated with photographic portraits is disrupted. The rhythm of the editing aligns with the music, so that stand-out lyrics are accompanied with expressive faces.

FIGURE 5.1   *Orange is the New Black* title sequence. One portrait in a sequence of images focused on the eyes. All frame grabs are available on Netflix's YouTube channel.

As we've discussed in other chapters, the direct gaze can be an invitation to make an emotional contact or alternatively it can be confrontational or accusatory (Lutz and Collins, 1993: 176, 197). Along with the distance and angle, the gaze creates 'a visual form of direct address' demanding 'that the viewer enter into some kind of imaginary relation with him or her' (Kress and van Leeuwen, 2006: 122). The gaze is not only about the act of looking, but about social relationships in a particular set of circumstances. In this case, the extreme close-up serves to humanize rather than objectify, and there is also a hint of sensuous sexuality in the focus on the lips (they are close enough to kiss). The movement within the pictures also adds to the sense of social interaction as the women smile or laugh back (Figure 5.2).

It is useful here to connect with writing on the photographic portrait, while recognizing that the sequence is a televisual format. Photography theorist Graham Clarke explains how portraiture has strong if problematic associations with identity and authenticity: 'the portrait photograph surreptitiously declares itself as the *trace* of the person (or personality) before the eye' (Clarke, 1992: 1, emphasis in original). But while declaring itself as an authentic 'trace' or presence of the individual, there is an enigma in what the portrait crops out – its promise of intimately revealing something of the individual's personality is deceptive: 'It consistently offers the promise of the individual through a

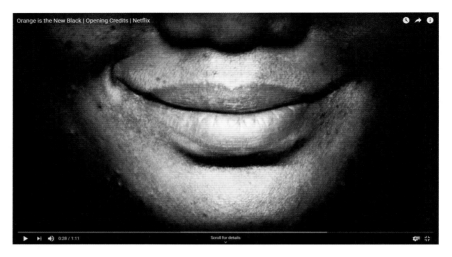

FIGURE 5.2 The lips and smiles of the women also feature. *OITNB* opening credits.

system of representation which at once hides and distorts the subject before the lens' (1992: 3). For Clarke, this paradox is at the centre of the portrait image, its 'compulsive ambiguity' promising both the traces of individual identity while displacing them with a necessarily selective representation.

This paradox of essence and enigma, of intimacy and anonymity, is nicely captured in the title sequence. In this context, the portraits are edited together so quickly and the faces are so closely framed that the sequence appears to capture the dialectical relationship between individualism and universalism that characterizes the humanist paradigm in portraiture (Lutz and Collins, 1993: 97). The portraits might offer authentic 'traces' of personalities but in a highly decontextualized and anonymized manner. There is no time here to scrutinize the faces or to find out more about their personal history.

The sequence is bold and distinctive. It could be argued that the repetitive style of closely cropped faces recalls the 'mug-shot' image and so conjures notions of categorization and institutional subjectification. Are the facial piercings and tattoos signifiers for criminality or gang culture? Our own interpretation suggests not, but others could, of course, read this differently. Instead, we would argue that the title sequence is effective in conveying the *ensemble* nature of the series and in foregrounding *diversity* as part of its self-branding.

Case study conclusion: the title sequence as a taster for telling diverse stories

The title sequence is a small if significant part of the show, and we have not addressed the series' main characters or storylines here. There are clearly avenues to extend this **visual discourse analysis** further into the ways in which the show represents the injustices of prison life, lesbianism, transphobia, criminality, institutional racism or the 'visibility of blackness' (Farr, 2016), to name a few possibilities. For instance, Brittany Farr credits the show for creating sympathetic portraits of incarcerated women but believes its humanizing approach can work to obscure the power dynamics operating within the prison 'providing the cachet of diversity without having to ask hard questions' (Farr, 2016: 158).

The show's representation of 'queer desire' is also a prominent theme, with San Filippo (2017: 86) commenting on the positive foregrounding of 'romantic-erotic attachments between women' in the series, while occasionally falling back on prison tropes of 'predatory lesbians and mercenary bisexuals'. For Farr (2016: 167), it is the activism of transgender actor Laverne Cox which is the show's most important success, demonstrating how 'visibility and representation can reach beyond the representational sphere' in Cox's commitment to 'highlighting the violence and structural inequalities faced by transwomen of color in press interviews'. There is undoubtedly a great deal more to examine about this series, especially in relation to how it 'packages' diversity in its self-branding and narrative themes, along with its relevance to the politics of representation beyond the show's own storylines.

FOCUS

THE EUROPEAN REFUGEE 'CRISIS'

The European refugee 'crisis' of 2015 became labelled as such because those fleeing wars and failed states in Syria, Afghanistan and Libya, among others, were reaching Europe's borders in ever larger numbers (increasing rapidly to up to around 1 million refugees that year). The sudden increase became a 'crisis' moment for European leaders as a combination of militarized borders and divergent policies on migration across the nation states of the EU exposed divisions and led to further hardening of policies to keep migrants and refugees out.

A number of journal articles and special issues have been written in response to the crisis with a specific focus on how the media have reported on refugees and migrants, and what the effects of their reporting choices might be, for both public opinion and policy. Those focusing specifically on visual representation, whether in news 'framing' or on social media, highlight the prevalence of dominant stereotypical patterns which work to objectify, collectivize, and thus dehumanize immigrants and refugees (see Smets and Bozdağ, 2018 for a recent overview and editorial introduction).

(Continued)

> In their study of headline images across five European countries, Lilie Chouliaraki and Tijana Stolić (2017) develop a typology of refugee visibilities, and question whether and how the visual regime encourages civic agency and a responsibility to act in the face of suffering. Chouliaraki and Stolić (2017: 1173) identify strategies of dehumanization in news images, casting the refugee as 'a body-in-need, a powerless child, a racial "other", and linguistic token or a sentimental drawing', but they also propose alternative ways in which the lives of refugees can appear as acting with us 'in a common world', telling their own stories. In this proposed visual regime, they appear in images that they have photographed themselves; through actions that portray them as creative and knowledgeable actors rather than as victims or terrorists; as citizens with views on the causes of the 'crisis' and as professionals with ideas and aspirations (2017: 1174).
>
> The above study provides a useful summary of both the problematic tropes of photography that work to construct boundaries between 'us' and 'them', and the types of imagery that offer the possibility of recovering human dignity, in the representational sphere at least.

CASE STUDY 2: HUMANIZING THE OTHER: SAVE THE CHILDREN'S 'REMEMBER ROHINGYA' ADVERT CAMPAIGN

In 2019, Save the Children celebrated its centenary. Founded in the wake of the First World War to help refugee children facing famine across Europe, the British charity is now considered one of the 'big' international non-governmental organizations (NGOs), or 'BINGOs', with others including Oxfam and Médecins Sans Frontières (MSF). Their media campaigns have attracted scholarly attention, in some cases analysed alongside the ethical guidelines and codes of conduct drawn up to help NGOs avoid images that 'foster a sense of Northern superiority' ('Code of Conduct on Images and Messages' cited in Manzo, 2008: 637). With its established focus on 'saving children', Save the Children has come under fire for using 'shock' images of starving children looking pleadingly into the camera (and therefore at their potential saviours), in a manner that reinforces a paternal logic with traces of colonial iconography (Manzo, 2008: 636; see also Zarzycka, 2016). Children are the ultimate emblems of innocent vulnerability and dependence, with pictures of their faces employed as a principal way for charities to attract donations and support.

Charities face a double bind: they use images of innocent suffering to capture empathetic attention and to encourage donations or petition-signing, but the same images stand accused of robbing those pictured of agency and dignity. Kate Manzo writes that the continued reliance on such imagery is paradoxical rather than unprincipled: 'For all its faults, the "starving baby" image *is* a powerful icon of human suffering thanks to the cultural connotations attached to its compositional elements' (2008: 638). As Manzo explains, the associations of the signifiers of famine have been constructed intertextually in relation to previous media appeals, and so the connotation of impending death is 'logically consistent with the humanitarian imperative and the injunction to save life' (2008: 639).

In grappling with this paradox, humanitarian charities have turned to different visual strategies – posting success stories and images of children smiling and being helped by charity workers. Children are not pictured alone but with adults who are providing food, education or medical aid. But this also holds resonances of colonial 'rescue' and protection when those pictured helping are not local staff or partners. Manzo (2008) finds the most promising instances in campaigns where NGOs shift away from a position of neutrality and instead make political demands of the reader or viewer underpinned by the notions of solidarity and justice rather than dependence.

Charities such as Save the Children are more than aware of the tensions involved in negotiating their objectives to draw attention to the vulnerability of those suffering, to follow ethical practices that protect human dignity and to avoid a reaffirmation of the colonial gaze. Working in conflict zones and humanitarian emergency situations, NGOs are well placed to provide pictures and stories for wider media distribution, especially in the case of overlooked or less visible crisis zones. In the contemporary media ecology, NGOs work closely with photojournalists and digital media experts to produce creative, unconventional and humanizing visual storytelling (Dencik and Allan, 2017).

Research questions and approach

To examine the ways in which Save the Children deal with the paradoxes and tensions outlined above, we turn now to a 2018 campaign calling for justice for Rohingya refugee children. Save the Children's media content productivity is staggering, and their website and social media activity alone provide a wealth of material covering various emergency appeals alongside child protection and education work. With visual communication in mind, we are particularly interested in how Save the Children employ visual discourse strategies in line with their own guidance and which also tell engaging stories.

It is not only academics who have a research interest in the images produced by Save the Children. In addition to updating their own guidelines regularly, in 2017 the charity published its own research report, 'The People in the Pictures', based on their 2014 four-country (UK, Jordan, Bangladesh and Niger) fieldwork conducted 'to listen to and learn from those who contribute their images and stories' (Warrington and Crombie, 2017). The report summarizes interview data about the contributors' and staff experiences of the image-making process, and focuses on issues of consent, agency and accountability. One of the recommendations is to 'invest in creative and collaborative approaches to image making', and within this remit they consider the value of participatory media projects and recognizing children as spokespeople for the wider contextual issues affecting them, not only a voice for their own personalized story.

With the report's research and recommendations in mind, we consider the **visual discourse strategy** employed in the Save the Children appeal for the Rohingya refugees. We ask the following questions:

- In what ways does the Save the Children appeal for the Rohingya refugees enact the recommendations of the charity's own guidelines and address concerns about ethical representations of children?
- How is the imperative to humanize the children's plight realized through specific visual discourse strategies?

Findings: how Save the Children's 'Remember Rohingya' campaign employs creative visual strategies

We first spotted this advert in the newspaper in August 2018, headlined 'Justice of Rohingya children' with a link to a petition designed to persuade the UK Foreign Secretary, Jeremy Hunt, to refer Myanmar to the International Criminal Court. To offer a very brief account, the Rohingya people are a Muslim minority group who have been denied citizenship in Myanmar despite around a million Rohingya living in the country. Following earlier communal violence, the numbers of displaced people rose drastically in 2017, when around 700,000 people fled Myanmar for neighbouring Bangladesh following the destruction of their homes, killings and rape carried out by the Myanmar security forces. The Myanmar government disputes the claims of the Rohingya, counter-claiming that the villages were destroyed by militants. Despite an agreement to repatriate refugees, many are fearful of further oppression and remain in camps vulnerable to torrential rain and disease.

Returning to the Save the Children advert, the stated purpose is political and legal – to bring the perpetrators of the violence to justice via international law. The notable difference is the visual image: not a photograph but a drawing that depicts armed men shooting children with a helicopter hovering in the background. The caption asks: 'What drives a child to draw a picture like this?'

The newspaper advert is just one element in the campaign, and a video collates more of the drawings together with a soundtrack, available on the Save the Children YouTube channel (www.youtube.com/watch?v=OoFZIkbMOAI). In the video, the drawings are animated, with the voices of the children who made the drawings recounting what happened in their own language. A caption reads 'Voice of Majuma*, 12' (the asterisk indicates this is not her real name).

FIGURE 5.3 Save the Children's 'Remember Rohingya' campaign advert. Captured from the NGO's YouTube channel.

The spoken words are translated into English on the screen, written in childlike lettering (see Figure 5.3). In addition to the animated drawings, the screen fades to black before showing the child artist holding up the original drawing, her identity still protected by the face being cropped out of the video (Figure 5.4). The subtitles read: 'After I am educated, I want to be a teacher and I want to teach children.'

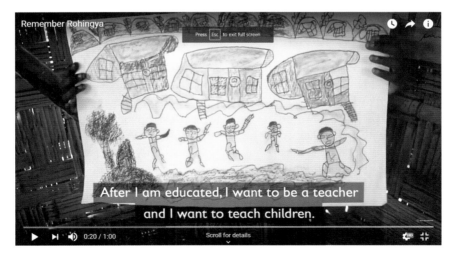

FIGURE 5.4 Save the Children's 'Remember Rohingya' campaign advert. An unidentified child holds up her drawing.

After fading to black and then to white, a second animation starts, with the voice of 'Mohammed*, 10' recounting how his house was burned down and they were driven out by men with guns (Figure 5.5). According to the subtitles, he adds: 'We don't want to spend our whole lives as refugees.' In both cases, the soundtrack adds a scribbling sound effect to connote the act of drawing, with minimal music and additional sounds to signify the surrounding environment (running river water), the quick breath of anxious running away, the gun shots and helicopter.

FIGURE 5.5 Save the Children's 'Remember Rohingya' campaign. Mohammed's drawing depicts men with guns burning down their homes.

A news article about the campaign on *The Independent* website reports how Save the Children had asked Rohingya children in the camps 'to think about a message they would like to send world leaders through their drawings or paintings' (Osborne, 2018). The news article includes further drawings and photographs of refugees with their identity protected, in addition to embedding the video. Contrary to the dominant trope of innocent vulnerability discussed above, there are no children's faces in any of these pictures. The news report is itself part of the intertextual humanitarian meta-discourse within which the appeal circulates, with extensive quotes from George Graham, Save the Children's Director of Humanitarian Policy, Advocacy and Campaigns, and Unicef spokesman Simon Ingram. *The Independent* web article is an example of the news media and NGO working closely together to amplify the message of 'remembering Rohingya' one year on from the attacks.

Finding new ways to visualize the vulnerability and innocence of children

The Rohingya appeal conforms to the recommendations mentioned earlier resulting from Save the Children's own research into their use of image and stories: that of investing 'in creative and collaborative approaches to image making' and recognizing children as spokespeople for the wider contextual issues affecting them. The appeal realizes this in a number of ways:

- Giving voice to the children who talk of both past oppression but also their hopes for the future.
- Using participatory practices to prompt recollections.
- Protecting the anonymity of the children through use of drawings and closely cropped video that omits the face.
- Focusing on justice for the children and so underlining the political accountability of the international community as well as the criminal culpability of Myanmar.

The decision to use children's drawings as the main visual strategy appears at first to abandon the evidential quality of photography. But this is not the full story. The children's own recorded voices and the transition from the animation to the film of the child holding up the drawing places the animation in the 'real world' of the refugee camp. The decontextualized drawings that offered few lifelike details or visual depth are now placed within naturalistic settings where small fingers grasp the sides of the paper in what looks like a dark makeshift shelter. The omission of the faces works to enhance the authenticity of the account by visually signaling the need to protect the identity of the children. In a similar manner, the two children are named in a personalizing manner (Majuma and Mohammed), but the names are not 'real'.

The child's drawing embeds their quality of innocence without having to show a pleading face. There is an 'iconography of childhood' (Manzo, 2008) deployed here, but with the child as artist and owner of their story. Yet the innocence inscribed in their style of drawing contrasts sharply with the scenes of killing and torture they depict and describe. The dialectic here between (innocent) style and (violent) content produces a disconcerting effect – an effect that the charity hopes will lead to a sense of responsibility and political action in the viewer.

Case study conclusion: imag(in)ing refugees as creative and knowledgeable

In Chouliaraki and Stolić's proposed visual regime in which refugees appear as acting with us 'in a common world', they suggest that refugees 'appear in images that they have photographed themselves; through actions that portray them as creative and knowledgeable actors' (2017: 1174; see Focus box in this chapter). While this example is NGO material rather than news images, the Save the Children appeal certainly seems to be creating the opportunities for such imagery to emerge.

The drawings are reminiscent of creative and 'participatory' research methods applied in visual anthropology and visual sociology where visual materials are used to elicit information and emotions (Pauwels, 2015a). Psychologists have long been interested in what **children's drawings** reveal about their development and social contexts: a particular interest has emerged in children's drawings of war and how the drawing process provides insights into their traumatic experiences. Indeed, various art therapy initiatives with Syrian refugee children have garnered news coverage in recent years. In all cases, whether for research purposes or as art therapy conducted by NGOs, the image-making process itself is valued and taken seriously, as a technique for overcoming trauma and/or a form of creative expression. In the Save the Children 'Remember Rohingya' appeal, the drawings are repurposed and reanimated, with the intention of bringing about change for those children. We cannot yet know if this particular campaign had any success in achieving its stated goals, but its innovative approach has at least demonstrated an alternative visual strategy to the overly familiar tropes, one that captures the creative agency of the children affected.

This is only a single appeal in the vast amount of campaign material produced by Save the Children, and we have admittedly selected it for its break with more familiar visual conventions. A quick scan of the charity's website reveals a range of images, including the lone child with the pleading gaze, but this type of image is, in fact, one among a diverse selection. For some, the uneasiness prompted by any NGO imagery is connected to the sense that it legitimizes an underlying notion of Western superiority entwined with a development agenda in which 'we' are best at protecting 'them'. As visual communication scholars, the crucial tasks ahead include continued vigilance in terms of monitoring the ethical nature of image-making processes, and being watchful that portrayals of humanitarian crises balance emotive bodies-in-need with humanizing agency and dignity.

CHAPTER CONCLUSION: ONGOING QUESTIONS REGARDING THE IDEOLOGICAL WORK OF VISUAL IMAGES

This chapter has explored the harmful potential of stereotypical representations and also demonstrated how keen attention to visual discourse practices and techniques can reveal the ideological work underpinning the choices made in diverse forms of media culture. Our case studies were designed to show how visual discourse strategies are employed by producers in an effort to disturb familiar representations or power relations, rather than assuming a 'one size fits all' (Weintraub, 2009) approach to discourse analysis. As David Weintraub (2009) advises in his useful framework, a meaningful discourse analysis not only looks for ideologies of capitalism and imperialism, but asks subtle and nuanced questions of the visual media images under investigation.

It is useful, then, to think of media representations as a series of **choices**. Who is selected to represent or symbolize a cultural group or community? What visual markers or characteristics are emphasized? What settings and behaviours are depicted?

Relating the above to wider contextual questions: do these visual representations arise due to producer biases or institutional pressures (or both)? How are power relations inscribed into familiar stereotypes? How do historical and contemporary regimes of power (re)produce and normalize 'truths' about certain groups? How do social attitudes and production cultures influence such patterns of representations over time? And how do later adjustments potentially shape attitudes, and even policies or laws? What kind of spectator is imagined in the form of address?

These are familiar questions to those engaged in critical discourse studies, but even the most familiar questions can remind us to question the way we encounter others in mediated forms, especially when their lived experiences would otherwise remain unseen and unknown. The next section of the book turns to our thematic strand of 'politics' where we continue to probe issues related to the politics of representation while paying specific attention to political imagery.

FURTHER READING

Stuart Hall's *Representation* (1997) and Kathryn Woodward's *Identity and Difference* (1997) remain excellent introductory guides to the themes discussed in this chapter, while the updated *Practices of Looking* (2017) by Marita Sturken and Lisa Cartwright provides topical examples and is more specifically focused on visual imagery. For a valuable question-led framework on discourse analysis of visual images, see David Weintraub's chapter in Keith Kenney's *Visual Communication Research Designs* (2009).

PART II
POLITICS

6 Images of politicians in the public sphere 109

7 The visual spectacles of protest and activism 135

8 Picturing international conflict and war 159

6 IMAGES OF POLITICIANS IN THE PUBLIC SPHERE

The next three chapters come under our general heading of 'politics'. We explore how politics is embedded in forms of representation, starting with politicians and political campaigning, but also embracing political activism (Chapter 7) and international conflict (Chapter 8). In each sphere, visual images perform influential roles.

Political leaders are now expected to perform on diverse political stages, adapting their personas to the affordances of each media genre and platform (Corner and Pels, 2003; Richardson et al., 2012). In this chapter we consider the fragile nature of 'mediated visibility' for politicians observed in traditional media coverage, as well as new concerns about the manipulative power of images across social media platforms.

In sum, this chapter:

- explains why traditionally there has been an uneasy relationship between the political realm and the visual;
- explores how politicians have sought to control their image and visibility in varied media forms and how digital technologies offer both potential opportunities and pitfalls;
- examines how politics is gendered in its visual representations and what this means for female politicians;
- conducts an analysis of the 'visual cues' found in the gendered representation of female politicians, with a focus on front-page newspaper covers of Theresa May during Brexit negotiations (Case study 1);
- looks to the dark side of meme culture and virality in an analysis of the visual rhetoric of disinformation found in the Russia-linked Facebook adverts from the 2016 US presidential election (Case study 2).

In the following sections we outline the theoretical context for thinking about the visual dimension in political communication, and how this is intertwined with notions of the spectacle, performance and symbolism.

THE TRADITIONAL SUSPICION OF THE VISUAL IMAGE IN POLITICAL COMMUNICATION

It goes without saying, perhaps, that we as visual communication researchers think mediated images play a significant role in constituting how events are perceived, affectively experienced and understood. But, as we have written elsewhere (Aiello and Parry, 2015; Parry, 2015), the long-held suspicion expressed towards imagery (manipulation, stylization, abstraction) reveals a sense of the 'flawed' nature of visual representation.

As television intertwined with the dominant mode of political campaigning and became a main source for public knowledge during the second half of the twentieth century, fears of a distracted and passive public or citizenry were associated with the particular qualities of the medium. Concerns were raised over the diminishing of political life into a spectacle, distorted by a media-driven shift that promoted conflict, sensationalism and inauthentic celebrity politicians. Politicians no longer had to concentrate on oratory skills designed to reach the back of the meeting hall; instead, they were required to finesse their authentic and intimate style for a TV close-up.

Underpinning the anxieties summarized above is the broader concern about the role of media in constituting the **public sphere**. Put simply, the public sphere is the space where citizens debate issues, freely exchange opinions and form their political identities. Key theorists of the public sphere, such as Jürgen Habermas, emphasize meaningful political participation in terms of 'voice' and rational deliberation or dialogue. In so doing, a form of iconoclasm or 'iconophobia' persists in the repudiation of images and vision as significant in the forming and shaping of publics (Finnegan and Kang, 2004). Crudely expressed, where talk leads to meaningful political action, images mystify and distract (Aiello and Parry, 2015).

Research has shown, however, that political information comes from a variety of sources, including comedy, political fiction, blogs and satirical websites, and so political realities are shaped through diverse ways of interacting with politics that go far beyond the news bulletins and political commentary newspaper pages. Where some authors see politics tarnished by the blurring lines of political information and entertainment values, others see an enhancement for democratic life in the popular engagement with politics encouraged by a variety of formats offering a mix of serious and more playful generic recipes (Corner and Pels, 2003; van Zoonen, 2005; Richardson et al., 2012). And with politics performed across such diverse media platforms, the visual dimension is often a significant part of the formula.

While the suspicion towards the 'visual' and 'affective' dimensions of politics continues to underpin concerns about political spectacle, celebrity politics and emotional publics, there is at least growing empirical and theoretical attention to the array of functions that visual images and symbols perform in political communication (Schill, 2012; Barnhurst and Quinn, 2012). We now see much more 'room' in politically focused scholarship for the affective, cultural and aesthetic forms of communication as well as space for marginalized groups (see also Chapter 7).

In his extensive review of the literature on the political image and visual image, Dan Schill (2012) outlines ten functions of visuals in politics: argumentation/persuasion; agenda-setting; dramatization; emotion; image-building (of candidates' characters); identification (between politician and audience); documentation (evidential); societal symbol (flags, military, sport);

transportation (back in time or to an imagined future); and ambiguity (suggesting without saying). Schill's list provides a useful overview and highlights the complexities facing visual communication researchers, not least in how to evaluate the normative implications when mediated images are placed centre-stage in democratic practices: 'Not only are we now campaigning largely by pictures – we are also governing by pictures and these televised images create the political culture from which we debate candidates and policies' (2012: 133). Schill challenges the notion that images have a trivial or limited importance in politics, but he agrees it is no simple task to understand whether visual symbols are 'better or worse' for the public sphere or to evaluate their role in making better political decisions.

We are, then, starting to see visually focused studies pushing back against the linguistic primacy that implicitly holds the 'emotive' and 'aesthetic' image as subordinate to the 'rational' and 'deliberative' word. We would share some of the anxieties surrounding the quality of public service media and of public debate in the current political climate, but we would argue that this makes it ever more necessary to be attentive to the varied ways in which visual images contribute to public communication in a diversified media environment.

POLITICAL LEADERSHIP AND MEDIA VISIBILITY

Politicians rely on the media to make themselves visible to their electorates. Traditionally, this would occur via news coverage and current affairs or 'political talk' programmes. But appearances across popular media as diverse as soap operas, satirical comedy and chat shows provide politicians with new platforms for a more intimate or informal mode of address. Such opportunities have also extended to online media, where political candidates and parties routinely campaign using Facebook, Instagram and Twitter accounts, where they can mobilize the faithful and amplify support via digitally mediated social networks and participatory culture practices. Indeed, by harnessing the creativity and sharing practices of supporters, political parties can save money previously spent on expensive advertising consultants.

But with such opportunities come risks. As John B. Thompson argues (2005: 42), this 'new visibility' can be characterized as a double-edged sword for politicians: a 'source of a new and distinctive kind of *fragility*' in a more complex and less controllable information environment, and where public–private boundaries are rewritten. So where television and later online campaigning offer a sense of intimacy and personalized address, the relationship between power and visibility becomes more complex and unpredictable: 'given the nature of the Internet, it is much more difficult to control the flow of symbolic content within it, and hence much more difficult for those in power to ensure that the images made available to individuals are those they would wish to see circulated' (Thompson, 2005: 38).

Thompson here mentions 'symbolic content' and 'images', and we would agree that visual content should be included when studying the contours of 'visibility'. However, it is worth pointing out that not all political communication studies of 'media visibility' or 'leader image' actually analyse visual content in detail. Media visibility can be used to refer to the number of paragraphs or stories concerning political leaders, or the degree to which they are quoted or appear on screen. Similarly, concerns with 'leader image' and 'image management' do not necessarily entail an empirical focus on visual appearance but rather on the projection or evaluation of character traits such as competence and integrity. In their article on the ways in which former Australian prime minister Julia Gillard's leadership was delegitimized in the national press, Jack Holland and Katharine A.M. Wright (2017) employ a 'gender-sensitive discourse analysis' to explore how Gillard could neither be accepted as having the qualities of a leader, nor as a true 'Australian', simply due to being a woman. This is an engaging article on identity narratives and Australian political culture, but it excludes any visual analysis of the photographs that originally accompanied the newspaper articles, due to the research being conducted using Lexis Nexis-acquired textual data only. How might the selected visuals have contributed to the gendered framings of Gillard identified in this study? If we accept images as a source of political information and symbolism, what kind of knowledge about the case might be missing here? Of course, we can also be criticized for emphasizing images over the text (where the substantive arguments, narratives and framings are indeed made), but our aim here is to signal how the long neglect of visual components is starting to end, and how research that integrates images into studies of political communication promises to further enrich such scholarship.

GLASS CEILINGS AND IRON LADIES: FEMALE POLITICIANS IN THE NEWS

We do not have the space here to fully cover the growing body of work on the media representation of female politicians. Such studies do, however, offer much food for thought on the ways in which political performance is gendered and how the news media work to reinforce certain expectations around female 'access' to the political domain, even if they do not necessarily examine how *visual images* contribute to this. For example, a number of studies demonstrate how the political realm is constructed as a masculinized territory, in which masculine notions of politics are normalized and female politicians are 'othered' as exceptions to the norm (van Zoonen, 2006). One way this happens is where female politicians come under scrutiny for being 'tough' or 'man' enough to be successful in the political arena (Harmer et al., 2017; Meeks, 2012), but also in the way in which male politicians are feminized as a means to discredit them (Parry and Richardson, 2011). Other gendered stereotypes are also found to work against women: with qualities of compassion and emotionality cast against decisive leadership traits; or female politicians sidelined in favour of the glamorous wives of male politicians (O'Neill et al., 2016).

As the above scholars and others have observed, female politicians face a double-bind: they struggle harder for media visibility, but when they do receive it, risk being mocked for their appearance, having their personal lives scrutinized, or become cast as neither a credible woman nor politician. There are, however, some signs of hope – a content analysis including six Western countries indicated that where progress is made in gender equality *at a societal level*, news coverage increases for female politicians (Humprecht and Esser, 2017). The flip-side to this is that progress in representational practices is only likely to occur in tandem with structural and societal changes. But as already mentioned, studies about the gendered nature of visibility and image do not necessarily devote attention to the actual visual portrayal (Wagner et al., 2017; O'Neill et al., 2016; Holland and Wright, 2017), but rather to verbal framings, stereotypes and metaphors, or the inclusion of direct quotations and certain gender-related themes. But there are a number of studies that are turning attention to political visual imagery and how the image and text work together (Dan and Iorgoveanu, 2013; Grabe and Bucy, 2009).

Visual analysis of the gendered portrayal of politicians

A key concern for female politicians is whether they receive a proportionate amount of coverage compared to male contemporaries, and whether that coverage is positively or negatively constructed. With a visual focus in mind, this might concentrate on the frequency and size of front-page newspapers or magazine images (or website equivalents). The tone or valence of visual coverage has also been explored, identifying certain attributes that contribute to a favourable or non-favourable portrayal. In their summary of content analyses of politicians' visual news portrayal, Katharina Lobinger and Cornelia Brantner (2015: 18) group the previously studied attributes into three clusters: '(1) visual imagery relating to human interaction; (2) visual imagery relating to photographic setting and appearance; and (3) visual imagery relating to photographic production values'.

In the first category above, researchers might consider non-verbal behaviour such as gesture, posture, eye contact, interactions with others (touching), facial expressions and attractiveness. The second set considers the depicted setting such as the place (Oval Office), who else is pictured (family members) and symbols such as flags. The final category refers to representational techniques '(such as camera angle, photo size, camera focus, light direction, light angle)' (Lobinger and

Brantner, 2015: 19). The clusters of visual cues outlined above cover the compositional and stylistic features of the photographs, but they refer only to the attributes of the visual imagery and not to the surrounding text or contextual factors that work in interplay with such portrayals. The three clusters do, however, offer a useful overview and possible framework for evaluating coverage of female and male politicians (extending to cross-country comparisons or in longitudinal studies). We return to this framework in our first case study later in the chapter.

POLITICAL ADVERTISING, VISUAL SOCIAL MEDIA, MEMES AND GIFS

Image-sharing has become central to communicative practices on social media, whether photographs, memes or animated GIFs (Graphics Interchange Format). This means that attention to political image management has shifted: to genres of political self-representation such as profile pictures on Facebook or 'selfies' on Instagram (Liebhart and Bernhardt, 2017), but also to the ways in which more traditional forms of political campaigning such as posters have moved online (Lee and Campbell, 2016). With the harnessing of social media platforms for political campaigning, there is a degree of continuity in the functions of the political visual image, as outlined by Schill earlier (2012) (e.g. image-building, emotional, identification), but there is also notable change, perhaps most clearly in the potential for interactivity (shares, likes, replies) which, even if in the most rudimentary way, capture forms of participatory culture and citizen engagement.

Images especially thrive in a digital world of mash-up, montage, juxtaposition, repetition and manipulation. The most popular internet memes, whether user-generated or popular culture-related, are humorous and playful but not necessarily political. Those dealing directly with politics can express a range of expressive modes, from light-hearted mockery to oppositional fervour. Studies of visual social media and politics have suffered from a familiar pattern, in the sense of lagging behind those studies focused on text. In part, this is due to the nature of the data accessible through the APIs (application programming interface) of platforms such as Twitter. As Tim Highfield and Tama Leaver (2016: 48) point out, '[t]he visual adds levels of trickiness to such analyses: first in accessing the images, videos, or other linked and embedded files, and then in studying them, which requires more individual intervention and interpretation than samples of 140-characters'.

Despite being an early version of a rudimentary form of digital animation dating back to 1987, GIFs have re-emerged on social media platforms as a way to express humour, irony and shared cultural reference points, and are currently especially popular in tweets. Kate Miltner and Tim Highfield (2017) set out how the particular features and affordances of GIFs mean that they are used for certain cultural practices that differ from other visual media. One point they make is that the GIF is a 'community-oriented' format and so 'the use of specific GIFs has the capacity to create in- and out-group boundaries' (2017: 4). Indeed, the fact that a visual format associated with humour and vernacular digital practices has been appropriated by political parties demonstrates a readiness to appeal to voters through affective and precariously polysemic means. GIFs might appear inconsequential but, as Miltner and Highfield argue, the 'symbolic complexity makes them an ideal tool for enhancing two core aspects of digital communication: the performance of affect and the demonstration of cultural knowledge' (2017: 1). Politicians ignore such capabilities at their peril.

Whether it is image-macros with their distinguishable 'Impact' font, parody accounts, online image generators (where various visual or textual messages can be inserted into a recognizable image-frame to imitate a genuine photograph or video clip), or reaction GIFs, these forms of memetic media are part of the public conversations generated in response to political events (as well as popular cultural moments). Ryan Milner argues in his book *The World Made Meme* (2016) that there is potential here for an empowering multiplicity of voices in what remains an undeniably unequal online ecosystem, but also space for antagonism and hostility to thrive. But where memes overtly signal their constructed nature, creativity and playful expression, the circulation of deliberately *false* information during recent elections, often coordinated in trolling attacks or even by automated bots, has raised deep concerns about the ways in which public conversations online are open to distortion and how democratic outcomes might even be affected.

New concerns for the misleading and manipulative image

Disinformation, especially that spread through online websites and social media services, has become a much discussed concern in recent years. In 2016, *Oxford Dictionaries* picked 'post-truth' as its 'word of the year', and in 2017, *Collins Dictionary* selected 'fake news', defined as 'false, often sensational, information disseminated under the guise of news reporting'. Many commentators and scholars have warned against the use of terms such as 'fake news' and 'post-truth' due to the ways in which such words set up unhelpful binaries and have become politically 'weaponized'. In an effort to combat this, we can note a number of national and supranational initiatives taking shape, not just through individual research articles, but in the formation of fact checking and debunking networks (such as the International Fact-Checking Network), open-access reports that demonstrate how to identify the sources of false and manipulative information (Bounegru et al., 2017), and in presentations to various congressional, parliamentary and European Commission enquiries from academic researchers, journalists and the 'big tech' companies (Wardle and Derakhshan, 2017).

Diverse forms of visual material act as persuasive resources for those who wish to attract 'clicks' and encourage others to share or refashion messages through online social networks. Photographs, memes, diagrams and charts add an emotionally appealing, forceful, symbolic and/or evidential quality to any mediated message, and so are not only ideal for genuine advertisers and political campaigners, but also for those hoping to use such qualities for subversive purposes. We return to this kind of material in our second case study to demonstrate how visual qualities and familiar tropes are deployed in fabricated social media posts. First, we examine how the visibility afforded to politicians in the wider media landscape can indeed be a double-edged sword, with female political leaders facing additional challenges.

CASE STUDY 1: THERESA MAY ON THE FRONT PAGE DURING BREXIT NEGOTIATIONS

Our first case study considers the visual coverage of British Prime Minister Theresa May during a tumultuous year of post-Brexit negotiations in 2017, as the United Kingdom (UK) held talks with European Union (EU) representatives. In the aftermath of the referendum result, Theresa May stepped into the prime ministerial role and therefore became the key political figure in determining what 'Brexit' might mean for the UK and how to negotiate its exit from the EU.

As other studies on politics and gender have suggested, political strategists and the news media each work to construct political actors as characters performing on a political stage or as '*person[s] of qualities*' (Corner, 2000: 393), put forward for appraisal by the public. Here we adopt an approach that primarily aims to show how this works visually rather than discursively. The linguistic content is here treated as secondary to the visuals in order to show how the images constitute the political character and 'political picture story' of Prime Minister Theresa May.

Research questions and approach

This case study examines UK newspaper digital front pages of Theresa May at significant points in the Brexit negotiations during 2017. The digitized newspaper front page crosses over from the old media to the new as a signifier for the most controversial, appealing or scandalous story of the day. In the UK *The Independent* newspaper continues to produce a 'front page' image even though it ceased publishing its print edition in March 2016. Indeed, a provocative front page could now be designed to imitate the viral quality of online images – it gets the news brand noticed, even when shared in apparent disgust. Our research questions are:

- Which newspapers choose to illustrate their front pages with a photograph or cartoon of Theresa May on these key dates?
- How do the visual cues such as appearance, setting and representational techniques combine to suggest certain character traits, personal qualities or power relations?
- Can we identify particular codes of a gendered representation?

The sample is selected by researching the dates for major meetings, speeches or votes in Parliament. The website Paperboy (www.thepaperboy.com/uk/front-pages.cfm) is used to find the relevant front pages for six daily newspapers that represent a broad spectrum of UK news media (tabloid and broadsheet; left and right wing; pro and anti-Brexit). The newspapers are the *Daily Express*, *Daily Mail*, *Daily Mirror*, *The Guardian*, *The Daily Telegraph* and *The Independent* (plus Sunday equivalents). We would like to have included *The Sun* and *The Times*, but the News Corporation newspapers are not accessible in this archive. With more space, a comparison with other European titles would also be interesting to pursue.

We focus on a selection from our sample and conduct an analysis based on the 'visual cues' outlined in the existing relevant literature by Lobinger and Brantner (see above, 2015). In sum, we consider:

- non-verbal behaviour such as gesture, posture, eye contact, interactions with others (touching), facial expressions, and attractiveness;
- the depicted setting such as the place (Parliament, Downing Street), who else is pictured (family members) and symbols such as flags;
- representational techniques: camera angle, photo size, depth of field, focus, lighting, distance.

We therefore focus our observations on the 'political picture story' of Theresa May offered by the visual images. Table 6.1 presents the key dates where we found at least three front pages with photographs or cartoons of May following major political developments. In the following findings section we pay particular attention to three of these dates: 18 January 2017, 28 March 2017 and 21 October 2017.

TABLE 6.1 Selection of dates for front-page coverage of Theresa May during 2017.

Date	Occasion	Paperboy hyperlink
18 Jan.	Lancaster House speech explaining how 'Brexit means Brexit'	www.thepaperboy.com/uk/2017/01/18/front-pages-archive.cfm
28 March	May meets with Nicola Sturgeon in Glasgow ahead of triggering Article 50	www.thepaperboy.com/uk/2017/03/28/front-pages-archive.cfm
22–23 June	European Council Summit	www.thepaperboy.com/uk/2017/06/23/front-pages-archive.cfm

Date	Occasion	Paperboy hyperlink
26 June	May £1 billion deal with DUP	www.thepaperboy.com/uk/2017/06/27/front-pages-archive.cfm
22 Sept.	Florence speech	www.thepaperboy.com/uk/2017/09/23/front-pages-archive.cfm
21 Oct.	May pictured alone at the EU negotiating table	www.thepaperboy.com/uk/2017/10/21/front-pages-archive.cfm
4 Dec.	May has to interrupt meeting to speak with DUP leader Arlene Foster	www.thepaperboy.com/uk/2017/12/05/front-pages-archive.cfm
9 Dec.	May and Juncker reach a dawn deal on stage 1	www.thepaperboy.com/uk/2017/12/09/front-pages-archive.cfm

Findings: from Iron Lady to lonely meme

As indicated above, this section focuses on the three categories of 'visual cues' as observed in a selection of front-page images rather than detailing the quantitative elements of the case study (i.e. number and size of images).

On 18 January, all six of the newspapers pictured May following her Lancaster House speech in which she indicated that Britain would leave the European Union with or without a deal in place. For those supportive of Brexit, this was a bold speech in which May laid down her ultimatum to EU leaders. Most newspapers followed a simple approach of a close-up head shot of May giving her speech. Two front pages stand out for their differing approaches. The *Daily Mail* chose to use a cartoon rather than a photograph, a visual tactic it reserves for special occasions, often accompanying an editorial – for example, during a general election campaign.

Printing a cartoon rather than a photograph allows the *Mail* to visually condense their support for May as a strong leader into a single picture and pack in a great deal of stark symbolism. May stands on the edge of the cliffs of Dover with her hands on her hips, handbag clutched at her side in a resolute stance. Given the headline 'Steel of the new Iron Lady', the handbag could also be a visual reference to Margaret Thatcher, the original Iron Lady renowned for always carrying her handbag. The *Mail* (along with other right-wing newspapers) often uses the white cliffs of Dover to represent a certain vision of Britain, or perhaps more correctly, England. This serves to highlight the country's geographical distance from Europe, across the Channel, as well as the political differences, crudely expressed with May actually standing upon the EU flag. In contrast, the Union Jack flutters in the breeze above May's head. There are implicit historical resonances here too. The symbolic vision of the UK courageously 'standing alone' brings to mind the persistent fascination with the Second World War in British political culture and the ways in which British identity continues to be constructed in relation to certain myths and pride related to this era. There is not a great deal of ambiguity here.

FIGURE 6.1 *Daily Mail* front page on 18 January 2017. All front pages can be accessed via the Paperboy website (see Table 6.1 for details).

In contrast, *The Independent* printed an extreme close-up of May with her eyes closed or looking down (see Table 6.1 for links to front-page images on 18 January 2017). While close-ups are generally thought to convey intimacy, a large size, *extreme* close-up (where the shoulders and even part of the face or head are omitted) becomes more intrusive (Kress and van Leeuwen, 2006), possibly probing the psyche of the subject or suggesting an insight into their inner thoughts. Printed across the top half of the page, this prominence suggests an unwelcome closeness, invading May's personal space via the camera lens. The closed eyes reinforce this sense of vulnerability or exposure, as we get the 'up-close' view of someone who, in that captured moment, is unable to return the gaze.

We move on to what is possibly the most controversial front page of Theresa May's premiership. On 28 March 2017, the *Daily Mail* attracted huge criticism and mocking disdain for its 'Never mind Brexit, who won Legs-it!' top-page headline accompanying a photograph of Theresa May and Scottish National Party leader Nicola Sturgeon meeting in Glasgow ahead of the Prime Minister triggering Article 50 (starting the process of leaving the EU). The headline explicitly casts aside the issue of Brexit in favour of an image-inspired commentary on the shape of both party leaders' legs (see Figure 6.2).

FIGURE 6.2 *Daily Mail* front page on 28 March 2017.

The direct but low angle of the photo in the *Mail* draws attention to their legs, with both women's heads appearing comparatively small and distant. The effect is maximized by the way the photograph is embedded into the title banner, with the heads and May's toes breaking the square frame so that the women's heads appear almost pushed out of the image. This breaking of the frame is a device often used by (tabloid) newspapers, which can add a dynamic or dramatic element and liven up the layout.

While the image does not portray May or Sturgeon in an unfavourable light (some might even say it is a flattering image to accompany a 'light-hearted' comment piece), the concern is that two of the most senior female politicians in the UK are pictured in a way that emphasizes appearance over their discussions on important issues such as Brexit and a second referendum on Scottish independence. The headline also encourages readers to make a judgement on the shape of their legs (even if the exclamation mark should strictly be a question mark) and so further orients the agenda towards how prominent women should be judged on their appearance. Underlying this is a perception that women's legs are somehow out of place in political meetings, otherwise, why would they attract this kind of attention?

We skip forward now to 21 October 2017, to a photograph from a two-day summit of European leaders in Brussels. Unlike the 'Legs-it' photograph, this photograph by Geert Vanden Wijngaert appeared in reviews of the year as a more thoughtful and well-composed image. It speaks volumes to photographers who are experts in political 'optics'. As most leaders know, it is best to avoid being pictured alone at a summit, especially when your country is in an isolated negotiating position. As Figure 6.3 shows, Theresa May sits alone at an apparently empty table, appearing deep in thought, with her hands beneath the table and her shoulders slightly hunched. There are no meeting papers or other signs of activity or purpose. This could be a photograph depicting a reflective and thoughtful leader, but the context of Brexit permeates the image so that her aloneness connotes an excluded diplomatic positioning. The colours are also significant here – a palette of greys, beige and white could denote a peaceful scene, but there is a funereal and washed-out quality to the image, not especially helped by the decorative white flowers. The angle also accentuates the size of the conference table so that May appears slightly lost and almost childlike in its expanse.

Perhaps not surprisingly given the pro-Brexit stance of most of our newspaper sample, the photograph appears only on the front page of the anti-Brexit *Guardian* and *Independent* broadsheet newspapers. *The Independent* prints the photograph across the whole of the top half of the paper, accompanied with the headline 'Brexit isn't working' and linking the photo to a poll in which 76 per cent of people agree that the government's strategy is going badly. *The Guardian* takes a more personal and even sympathetic approach with the caption 'Nobody said it would be easy…'.

FIGURE 6.3   *The Guardian* front page on 21 October 2017.

As well as later appearing in photographic reviews of the year, the photograph by Geert Vanden Wijngaert was shared on Twitter with various playful captions and adapted as a meme known as 'Lonely Theresa May' (http://knowyourmeme.com/memes/lonely-theresa-may/photos). Without going too far off-track here, this example also speaks to the ways in which political cultures and digital cultures have become entwined. One version of the 'Lonely Theresa May' meme also references the more established 'This is Fine' meme of a cartoon dog in a burning house drawn by K.C. Green in his *Gunshow* web-comic, typically used as a reaction image to convey a sense of self-denial 'in the face of a hopeless situation' (http://knowyourmeme.com/memes/this-is-fine (Figure 6.4).

FIGURE 6.4 The 'Lonely Theresa May' meme based on Geert Vanden Wijngaert's photograph, digitally mixed with the 'This is Fine' cartoon reaction image. Uploaded by 'Political_LOL-Center' and accessed via the 'Know your Meme' website.

Producing and sharing such images are as much about indicating shared cultural knowledge and knowingness as they are about mocking May in particular. Memes such as this are emblematic of a more playful engagement with politics, interwoven within both the malleability of digital visual technologies and the community-building participatory practices of the internet (Shifman, 2014). However, briefly introducing this example demonstrates how easily the political leader image can become further unmoored (or decontextualized) from the anticipated official version of political activity and performance.

Case study conclusion: a difficult year in the political spotlight

By all accounts, Theresa May did not have a great political year, losing her parliamentary majority after calling a snap general election she was predicted to win by a landslide. The Brexit negotiations offered an opportunity to demonstrate diplomacy and leadership, but also presented huge challenges for the prime minister during a politically divisive period. The above case study is only a 'snapshot' account of the front-page visual coverage she received, but we hope to have pointed out a number of key features that highlight the roles of visuals in political storytelling and in conveying the 'qualities' of political leaders. We have also deliberately avoided using any specific terminology associated with a certain method here (such as semiotics), but instead have drawn upon a set of 'visual cues' in order to analyse primarily the visual imagery. We have related this to a number of notable themes identified in the literature on visual political communication: the gendering of political leaders and the naturalization of the political realm as a masculine environment; the association of politicians with potent symbols and national identity; the dangers of getting the 'political optics' wrong in the hybrid media ecosystem. Examining the gestures, expressions, settings and compositional features of the recurring photographs of political leaders provides an insight not only into those depicted but also something about our wider political culture.

ACTIVITY

CONTROLLING THE OPTICS

> Political leaders recognize that a visual image conveying positive traits related to their political or personal lives can enhance their political image. A number of politicians have followed Obama's lead in employing a personal photographer to promote the perfectly composed image, but this can also lead to criticism from press or agency photographers who are denied independent access to key events. Find an example where a politician has attempted to control the 'optics' of a situation. What are the characteristics of an effective example? What are the failings when such an attempt goes wrong? How do online communities then attempt to remix such images? What do these examples tell us about the values or ideals associated with political leaders and impression management?

CASE STUDY 2: FAKE POLITICAL ADVERTISING ON SOCIAL MEDIA

As noted earlier in the chapter, the complexities of online image-sharing cultures bring concerns and ambivalences, especially where the original source of the disseminated material is misleading or deceptive. One controversy that has attracted a great deal of commentary and speculation is the apparent involvement of Russian-sponsored trolls and fake accounts on Facebook, Instagram and Twitter during the 2016 US presidential election (with similar concerns about the 2016 Brexit vote in the UK). The idea of unknown actors from another country attempting to influence domestic voters raises serious issues of sovereignty and democracy, but rather than necessarily using persuasive messages supportive of a certain candidate (in this case Donald Trump), it appears that the Russian 'interference' was multipolar and varied, and so intended to disrupt and divide rather than necessarily inspire votes.

The political adverts analysed here are interesting to us precisely because they employ visual images based on their capacity to persuade through symbolic and evidential capacities, but they are in fact 'fake'. This means that their deceptive function adds another layer to their meanings in the context we see them in now – revealed as inauthentic and unreliable. As 'knock-off' imitations of political marketing, these apparent adverts and community-building posts subvert the iconography of political campaigns for hidden propagandist motives.

Jesse Daniels defines 'cloaked websites' as 'those published by individuals or groups that conceal authorship or feign legitimacy in order to deliberately disguise a hidden political agenda' (Daniels, 2009: 661; see also Farkas et al., 2017 for a more recent study). The visual images and symbols help to establish the desired legitimacy, and to attract clicks and shares, even where they appear crude or artless. There are a growing number of examples of coordinated social media accounts attempting to polarize debate, foster social division and amplify hostility (often towards minority groups) during election campaigns and in the aftermath of terrorist attacks, as well as in other crisis situations. Researchers and policy-makers are grappling with how to identify the sources for such impersonator accounts (before they disappear) and how to diminish the harm they potentially cause.

Of course, it is worth stressing that we cannot know whether these adverts were deemed to be particularly believable or persuasive by those they were targeted at, but their dual function (on the surface, as overt and direct persuasion or information about events, covertly as disruptive propaganda), presents a multilayered set of social media artefacts ripe for analysis.

Research questions and approach

As ostensibly political advertising, we are dealing here with multimodal texts designed to function as persuasive communication. Therefore, we conduct a rhetorical analysis. As we have already outlined in Chapter 2, **visual rhetoric** concerns the strategic use of symbols to 'elicit a certain kind of response in the addressee' (Paul Grice cited in Rampley, 2005: 137). How are particular communicative devices used to generate a certain response? Influenced by Roland Barthes' *Mythologies* (1972), visual rhetorical analysis is particularly focused on associations through metaphor, intertextual references, and the 'structures of social, economic and cultural relations and power' in which such imagery is enmeshed (Rampley, 2005: 135). Visual rhetoric recognizes how symbols are chosen to convey shared values or ideals, and to build communities – for example, the way that flags symbolize national identity. Symbols or visual elements can also be adapted or reappropriated to connect two events or communities together (intertextual references), offering a shorthand and easily recognizable motif or trope; possibly to raise awareness or mobilize supporters for a political campaign (also see Chapter 7).

Our analysis therefore focuses on some of the qualities we have sketched out above in relation to memetic media and visual rhetoric (symbols and tropes, stylistic conventions of the specific media, intertextuality).

- What rhetorical devices are used?
- What might be the intended response (in the addressee)?

We only have access to this selection because Facebook submitted a larger collection to the US House of Representatives' Intelligence Committee hearing, and this small handful was made available on the Committee's Democrats website (along with metadata including the targeting information and a list of 2,753 suspended Twitter accounts (HPSCI, 2017; see also Lecher, 2017)). Of course, the wider context is crucial to any interpretation we offer here. We *now* know that these are false adverts, claiming to be something they are not. Our task here is to show *how* the adverts were intended to work without raising concerns about their credibility, embedded within the formats and functions of online social networks. We also consider the elements that suggest their inauthenticity.

Findings: the dark arts of visual propaganda

Archive imagery and intertextuality

We concentrate here on some of the more visually interesting or surprising posts. One advert presented in the evidence bundle provoked particular surprise and came from a Facebook account that had accumulated a substantial following (with 360,000 likes on Facebook before it was suspended (Levin, 2017)). The 'Blacktivist' account name clearly signals its apparent intent (a portmanteau of 'black activist'). In this advert 'Blacktivist' uses an archive black and white image of the Black Panthers to link the current fight for racial justice to the historical struggles of the Black Panthers – here identifying both the US government and the KKK (Ku Klux Klan) as enemies to this struggle (see Figure 6.5).

The caption to the image contains grammatical errors and awkward phrasing which could be an attempt to use vernacular language, but with hindsight might also signal a non-native English speaker behind the message. The larger text box

FIGURE 6.5 Post from the fake 'Blacktivist' Facebook account (HPSCI, 2017).

appears to suggest that the KKK still exists due to the government's choice to dismantle the Black Panthers, but in contrast, not to outlaw the KKK. The post extolls the addressee to 'never forget' this rather convoluted and awkward statement. However, the photograph adds a solemnity to the posting, the men standing unsmiling in a line, wearing matching berets, badges and jackets to convey their unity in purpose. The use of an archival black-and-white image is an attempt to add gravitas, evidentiality and a sense of history to the occasion. This is an **intertextual** resource, with the intended meaning only realized by understanding the 'codes' in the photograph and the implied relationship between the archive image and the Facebook posting.

A political symbol is also embedded in the bottom right corner – the clenched fist breaking through chains (which also serves as the anonymous profile picture for the account). The photograph and symbol are designed to project an *identity* for Blacktivist and show *support* for the cause. This is all about signalling solidarity and in-group status to those with shared political ideals: 'Symbols allow members to help create cohesion among the group and to create distinctions between those who belong to the campaign and those outside the campaign' (Goodnow, 2006: 175). The raised fist is a long-standing symbol of resistance, associated with socialist campaigns as well as the Black Power salute. The additional element of the breaking chains signifies a rising up out of slavery or suppression, and so adds further resonance for the intended addressee.

According to Sam Levin writing in *The Guardian*, the Blacktivist account had already been queried by other community activists for attempting to organize unofficial marches in Baltimore, but others found it presented a convincing alternative narrative (Levin, 2017). As we can see, its use of archival images and symbols clearly played a key part in its rhetorical strategy.

The visual trope of X-ing out

The next example features a **visual trope** with a long history in propaganda. Visual tropes are recurrent elements that often hold iconographical significance. As with the use of the word 'trope' in literature or in music, the term signifies an agreed-upon meaning that is easily accessible and resonant in people's minds. It is, then, a rhetorical technique or convention which is instantly recognizable. In this case, the trope we're referring to is the 'X' of the crossing out over a face. As Richard Popp and Andrew Mendelson trace in their analysis of *Time* magazine front covers, the 'X-ing' over the face has been used in 1945 (Hitler), in 2003 (Saddam Hussein) and in 2006 (Abu Musab al-Zarqawi), in each case signifying the apparent defeat or death of each leader. As they point out, the **intertextuality** here is important, as is the visually visceral nature of the trope which imposes some kind of connection between the evils of each subject: 'while *Time* would probably never verbally argue the equivalence of Hitler, Hussein, and al-Zarqawi' (Popp and Mendelson, 2010: 216). In other words, the visual image makes an argument that might be considered unsayable in words. Following the publication of Popp and Mendelson's article, the 'X-ing' trope was perhaps rather predictably used once again by *Time* to announce the death of Osama bin Laden in May 2011.

The 'X-ing' in the Facebook adverts is slightly different from the blood-red dripping crosses of *Time* magazine. But they still suggest the almost violent erasure of their subjects and there is still a textured quality to the 'X' (Figure 6.6).

In Figure 6.6 we can note that almost the entire face of Hillary Clinton is covered with what has the appearance of thick black paint (albeit digitally produced). The textured brush stroke quality is important here: there is a violence to the size of the strokes in relation to Clinton's face, as if her identity is being wiped from the image. The background is also significant; the US flag is just discernible but darkened in a foreboding manner. The caption reads that 'Hillary Clinton is the co-author of Obama's anti-police and anti-Constitutional propaganda'. It's not entirely clear what this sweeping statement refers to, but it follows the propagandist's trick of labelling anything disagreeable as 'propaganda' (also note how Donald Trump uses 'fake news' in this way). Finally, it's worth mentioning the function of the post: to publicize an event taking place in New York. Such posts do not, then, confine their influence to digital life online, but mobilize and

FIGURE 6.6 Post from the fake 'Being Patriotic' account promoting a 'Down with Hillary' demonstration with her face X-ed out (HPSCI, 2017).

even organize offline gatherings of like-minded people (assuming the '45 people going' are not also fake accounts).

However, it is not only anti-Clinton gatherings promoted in this manner. Joining in with demonstrations held after Trump's electoral victory, a post by 'BM' apparently encourages followers to join a New York anti-Trump protest in Union Square (not pictured here). Again, the background darkness appears to envelop the New York street scene, while Trump's disembodied head is covered with a red X, this time with a slight transparency in the brushstroke texture. This event is apparently organized by 'BM' (a deliberately close moniker to 'BLM' – 'Black Lives Matter'), who has convinced over 16,700 people to say they are going. Mimicking many of the demonstrations organized around this time to express horror and repudiation of Trump as 'my president', the post mixes doom-laden fear of the future with quotidian appeals to bring 'signs, snacks, water!'

Mixing familiar memetic conventions and dehumanization

Our final example is indicative of a number of the posts we now know about, and there was a clear Islamophobic propensity in their rhetorical address. Figure 6.7 is just one example where pictures of women in burkas or niqabs are used to signify danger to the United States. The 'Stop A.I.' (Stop All Invaders) account has a clear anti-immigration message, but the posts target Muslim people specifically.

The question marks inserted over each covered face in the photograph hint at the menace of the unseen, despite the image clearly depicting three bare-footed women chatting as they walk. The conflation of Muslim dress codes with a 'security risk' and terrorism is highly inflammatory. The post also uses the recognizable Impact typeface used in image-macro memes (see Brideau and Berret (2014) for a brief history), and so stylistically draws on the familiar aesthetic of memetic media. It looks like a 'proper' meme because it fits such design expectations. Calls for burka bans are the kind of depressing low-grade, minimum-effort demands often made by far-right groups across the Western world, designed

FIGURE 6.7 Post from the fake 'Stop All Invaders' account on a proposed burka ban in the United States (HPSCI, 2017).

to appeal to wider publics beyond their core supporters. The 'Stop A.I.' account appears to hit a nerve here with 55,000 shares, but again it is impossible to know whether these 'marks' of interaction are entirely genuine.

The exact motivations for the creation of such accounts might never be known, but it is clear that the adverts play to existing societal divisions, and even stoke fears and prejudices. It is also clear that the visual dimension is an important and so far under-examined aspect to this intended manipulation.

Case study conclusion: misdirecting voters with visual and verbal violence

Analysing the above images gives us insight into the visual symbols, conventions and tropes the creators of the adverts were peddling. In terms of the visual content, one recurrent feature is the visceral nature or even symbolic violence in the choice of visual trope and symbolism, designed to provoke an emotional response. This is often reinforced through the aggressive verbal text. Another feature is the way in which the messages often attempt to influence actual political behaviours beyond 'liking and sharing', including attending demonstrations.

In terms of intended response, there are apparent efforts to foster group solidarity and belonging – as well as to identify the 'out-groups' and those who are not welcome. Some of these are not surprising – anti-Muslim, anti-immigration and anti-Clinton. But others, including those not discussed here, promote LGBT activism, Black Lives Matter and support for Hillary Clinton and Bernie Sanders. And yet it was all fabricated. Finally, we detect an absence of the playful humour or joy we would associate with meme culture. This, perhaps, is the clue that these are in fact 'sock puppets' or 'cloaked' accounts impersonating a variety of political identities and groups, and with subversive intentions.

The material we consider here works with such recognizable symbols and visual tropes of persuasion to *appear* as political adverts, both in the guise of officially sponsored political campaigns and as posts by activist groups. Crucially, in this case there is another layer to this strategic form of communication: an intended **misdirection** of the persuasive message, a multimodal sleight of hand designed to deceive, obfuscate and antagonize.

Part of the challenge for researchers is the way in which such advertising is targeted at certain users and therefore not necessarily available to researchers. As noted above, these are simply a small selection delivered to Congress for a hearing and posted on the Intelligence Committee webpages. Whether a money-making scheme for bored teenagers or a state-sanctioned conspiracy to distort online public discourse, the creators employed visual rhetoric – symbols, intertextuality, tropes – to appeal to emotions and mobilize support for political campaigns. However, whichever political goals are behind these and other 'fake' or 'cloaked' sites, the wider issue is that they potentially create a crisis in legitimacy, bringing uncertainty to all online content – what kind of visual and digital literacies do we need in order to judge what is authentic and what is false? More importantly, how might we encourage those who design such platforms towards creating spaces where disinformation is de-emphasized or buried rather than promoted, without resorting to censorship and knee-jerk forms of ineffective regulation? Such questions are beyond the scope of this chapter, but the nature of visual communication will need to be considered in any initiative to encourage more ethical and accountable online practices, and economic models, of the major media technology corporations.

CHAPTER CONCLUSION: ASSESSING POLITICAL IMAGERY IN THE PUBLIC SPHERE

We have focused our attention on the site of media texts in this chapter (as opposed to production or audience research), applying forms of visual analysis to a variety of political communication images. We have examined how fragile and punishing the nature of media visibility can be for political leaders, especially when facing the additional challenges of a masculinized environment as a female politician. The new concerns about manipulative political images have also been explored through a case of misdirected and deceptive political campaigning images.

The studies we cover here point to the complexity in assessing the role of political imagery in the public sphere and how, as one mode of communication, it interacts with other modes, primarily text but also visual design and news branding. Political campaign images are one communicative resource offering informative, affective and symbolic messages, but which are variously interpreted, accepted or dismissed along with other bits of information encountered in the mediascape, in addition to interpersonal interactions. But recognizing such complexities should not negate the importance of better understanding the role of visual communication in shaping political exchanges and experiences.

The uncertainties of turbulent political events appear to be chiming with growing misgivings about the way in which major media technologies and global companies are managed and regulated. We believe that paying close attention to political imagery does not preclude a more rigorous debate about the public sphere and democratic health; rather, it is a vital part of such considerations.

As politicians face global challenges such as climate change, international terrorism and intercontinental refugee deaths, what might serious political progress 'look' like? Can innovative political images be used to challenge the prevailing imaginary of populism and widespread divisive rhetoric? Politicians work with projections and expectations – not only 'How can I make this idea/policy work?'

but 'How can this idea/policy *be seen* to work?'. This is not just about the politics of visibility and visual images, but also the politics of vision, visuality and how to imagine a better world; and in Chapter 7 we explore these themes further by turning our attention to politics 'from below'.

FURTHER READING

A new edited collection on *Visual Political Communication* (Veneti et al., 2019) brings together theoretical, methodological and empirical essays by scholars writing in this area. For a journal-length summary of political science's squeamishness towards digital visual media, see Dean (2018).

7
THE VISUAL SPECTACLES OF PROTEST AND ACTIVISM

From holding up banners at marches, creating media-friendly spectacles with daring escapades, to the selfie-led campaigns on social media, modern-day activism creates varied and striking mediated images. It is hard to imagine an activist group or social movement planning a campaign without thinking about the images it produces and the ways such visuals circulate across multiple media platforms.

Shifting focus from 'top-down' to 'bottom-up' politics, this chapter considers how protesters harness the rhetorical power of the visual to fight the powerful. In order to gain visibility, activists and social movements often exploit the carnivalesque and theatrical to attract media attention, using various forms of 'media', from graffiti to YouTube videos.

In sum, this chapter:

- outlines research that deals with the power of imagery for protest and activism;
- explores three levels of visual communication in this mix: the visual materials used by activists during events; the mass media representation of protest and dissent; and the 'self-mediation' used by activists to raise awareness and mobilize support;
- conducts semiotic analysis of the visual resources produced and inspired by the masked protesters of Pussy Riot (Case study 1);
- examines how a Black Lives Matter 'visual icon' is co-created in shared Twitter images (Case study 2).

Recent global events from 2011 onwards (Occupy, Arab Spring, Ukraine crisis) have generated heated discussion of how various media are implicated within the planning and execution of dissent (Bennett and Segerberg, 2012), but the images generated and shared at the various levels of mediation are not always examined beyond the phenomenon of their distribution. We believe that a visual communication approach facilitates valuable scrutiny of the representational form and the content of such images alongside the social practices and technologies associated with such striking images of protest. It's not just the sharing that's important, but what's shared.

PROTEST IN THE MEDIA

There is a growing scholarly interest in social movements and protest, and especially in the study of how 'the media' are entwined with the organization, mobilization and amplification work involved in various campaigns for social change. As leading researchers connecting these two fields of study, Alice Mattoni and Emiliano Treré (2014) observe that this entwining was not always the case, with traditional approaches to social movements rarely paying attention to media and communication. They also note that where relevant studies exist, they tend to focus on a single medium such as radio or television, and concern themselves with mainstream media or alternative media, rather than the dynamics between different media forms. In addition to this 'one-medium bias', Mattoni and Treré find a contemporary parallel with a 'technological-fascination bias', where the latest technological platform becomes the sole focus for research (e.g. Twitter), which risks overlooking 'the connections between multiple technologies, actors, and their practices' (2014: 255).

From the perspective of activists, we can sum up two traditional concerns when it comes to the uses and role of media:

1. How to deal with traditional media outlets who are generally considered socially conservative and even hostile to oppositional politics of protest events.
2. How to make effective use of self-produced media to organize and mobilize support.

These concerns correspond to some degree with the large body of academic studies on how mainstream media report on protest actions and an emergent scholarly focus on how activists use media as a campaign tool. Indeed, the 'protest paradigm' refers to the tendency of mainstream media to marginalize and delegitimize protesters through the use of certain reporting conventions (focusing on violence while de-emphasizing the activists' origins and arguments) (see Chan and Lee, 1984; McLeod and Hertog, 1992). But many scholars are now questioning whether this reporting paradigm still holds in the twenty-first century, as the era of limited choice in news channels and newspapers is overtaken by an era of abundance enabled by digital communication technologies (Cottle and Lester, 2011), and as the distinctions between 'mainstream' and 'alternative' also become blurred. Of course, it would be foolish to claim that the communicative and ideological power of established media institutions has disappeared, but the transformations in the digital media landscape have clearly offered new opportunities to those who have previously struggled to get their voices heard. New conditions of media production and circulation have impacted upon traditional media forms in myriad ways, even if they have not replaced them.

Whether a local community group or global network, the distribution opportunities of the internet offer the *potential* to connect ideas and co-ordinate actions without requiring access to expensive resources or time-heavy organizational skills. The activists' self-produced videos uploaded to YouTube or Twitter can be shared without the filter of mainstream media, while eyewitness images often make it on to the evening news, and might even be crucial to contradicting official narratives of events. So the two concerns outlined above remain valid, but we would argue that such concerns have become more entangled and ambiguous, as protesters plan events with spectacular or humorous media coverage in mind, and the logics or ideologies of the 'mainstream media' turn out to be multidimensional rather than singular. And in the mediated attempts of activists to make and contest meanings, the visual elements are crucial.

IMAGE POLITICS AND PUBLIC SCREENS

In the introduction to *Image Politics in the Middle East* (2013: 1), former journalist and Middle Eastern politics specialist Lina Khatib argues that political struggle is an 'inherently visually productive process. [. . .] It is a struggle over presence, over visibility.' Where protest is encountered, there will generally be a sense of struggle, conflict or contestation; born of a motivation to proffer an alternative image, identity, reality or vision, and a hope that it resonates with a wider public. The Middle East and North African region has certainly endured its share of political conflict in recent years, but the uprisings known as the 'Arab Spring' in 2011 brought unprecedented global media attention to the 'people power' expressed through huge gatherings in protest camps, the toppling of statues, graffiti, cartoons, symbolic hand signals, Facebook pages, and many other artefacts and pictures.

We now know that the revolutionary fervour was met with repressive and sometimes brutal counter measures in countries such as Egypt, Bahrain and Syria, but in 2011 the Arab Spring appeared to inspire other movements for political, social and economic change who also coordinated camps and occupations of symbolic public spaces: the 15M movement in Spain and the Occupy movement that started in New York. But while protesters took inspiration from each other, in the expression of arguments for radical change and in the circulation of images facilitated by various media platforms, in each case the political circumstances and societal issues such as unemployment should be seen as the original impetus for the movements. In other words, it is important to examine how various media tools, among other factors, shape the dynamics and visuality of protest without assigning a simplistic causality through the use of terms such as the 'Facebook revolution' (see Lim, 2013).

Of course, the dramatic and spectacular nature of contentious politics has a longer history than perhaps imagined, and the influential work of American sociologist Charles Tilly on claims-making performances and repertoires looks at the forms through which ordinary people make collective claims, and how they shift over time and vary under different political regimes through the centuries (for example, see Tilly's posthumously published book *Contentious Performances* (2008)). As Doerr et al. (2014: 559) write, while Tilly did not consider visual information directly, the concept of *performance* has become central to highlighting the 'embodied practices of protest and the visual codes routinely used to display dissent'. More specifically, they argue that 'visual analysis provides potent tools to study the performance of emotions in movements' (Doerr et al., 2014: 559) due to ways in which the symbolic and iconic power of images efficiently stir passions, anger, outrage, irony or pride.

This desire for *visibility* through creating spectacle and striking images has been theorized by Kevin DeLuca in his work on *Image Politics* (1999) and by DeLuca and Jennifer Peeples (2002) in their appeal to take seriously the dissemination of images through media technologies such as television and the internet. The notion of an 'image event' interrogates the rhetorical power of images in public debate and refers to the spectacle-driven tactics used by activist groups to gain visibility in the media and to shake audiences awake. In their 2002 article on the WTO protests in Seattle, DeLuca and Peeples propose that in order to understand new conditions for rhetoric, politics and activism, the notion of the '**public sphere**' can be supplemented with the '**public screen**'. The orientation of the 'public screen' recognizes:

> that most, and the most important, public discussions take place via "screens" – television, computer, and the front page of newspapers. Further, it suggests that we cannot simply adopt the term "public sphere" and all it entails, a term indebted to orality and print, for the current screen age. The new term takes seriously the work of media theorists suggesting that new technologies introduce new forms of social organization and new modes of perception. (2002: 131)

As mentioned in the Introduction to this book, we are wary of general statements about being flooded with images, and image saturation, but DeLuca and Peeples's idea of a 'public discourse of images' (2002: 133) identifies the potentials for political agency and civic participation beyond the traditional emphasis on dialogue and voice. It is important to note that DeLuca and Peeples stress the potential for miscommunication and of multiple viewing publics, with mediated events open to multiple readings. But this orientation to discussions via public screens recognizes how visual images play a crucial part in such interactions.

To unpack these relationships between visual materials, mediated visibility and networked public culture further, we briefly distinguish three manifestations: the visuality of objects employed in dissent (visual materials used by activists during events); the media coverage of protest events; and the self-mediation activities of dissent.

Visual materials used by activists during events

Even before we think about processes of mediation generated by a protest event, visual markers are used to self-identify and to show collective solidarity with a cause – badges, coloured clothing, flags, symbols and banners provide often instantly recognizable codes, capable of crossing linguistic barriers in transnational movements. Even slogans which are linguistic in content are presented in certain aesthetic styles and so are also visually symbolic in nature. As political geographer and activist Paul Routledge (1997) has argued, resistance is 'imagineered' in both 'immediate' and 'media-ted' spaces. A focus on the *immediate* draws our attention to the local, to resistance as embodied and experiential, to the ***occupation of space*** (often of a disruptive character), and therefore to how the 'presence' of the protest group is arranged. For Pollyanna Ruiz (2013: 264), the ways in which different groups wear masks during protest can work to unsettle power structures:

> the refusal to be seen and categorized by the state is empowering in that it exposes, and then unsettles, the power dynamics that structure public space. [. . .] the image of the masked face has the potential to make the previously unseen majority visible and is therefore a purposeful form of presence.

Hiding one's face can be seen as sinister or threatening, but, as we also see in our case study on Pussy Riot, we would agree with Ruiz that masking the face can hold political and cultural power, and be understood as connoting collectivity and solidarity. This follows the work of Italian sociologist Alberto Melucci (1989), who claimed that the form that activism or resistance takes is itself a type of media or message, operating through registers and signs that challenge the dominant codes. The activists' actions themselves announce *another way of being*, often exposing contradictions through exaggeration or theatre.

Media representation of protest and dissent

As signalled by Paul Routledge's quotation above, as well as studying the role of the visual in the *immediate* experience of dissent, resistance is also '**media-ted**'. We already mentioned how, traditionally, the coverage of protest by news media is characterized by trivialization, marginalization and delegitimization. Those hoping to secure publicity for their cause adapt their strategies precisely to attract media attention; but creating attractive 'image events' through dramatic or theatrical spectacles, and especially through symbolic violence, simultaneously holds the danger of casting the movement as 'deviants' and outsiders.

This relates to one of the ways in which 'framing' theory has been applied in research on activism in order to look for patterns in the 'news framing' of protest. Framing analysis scrutinizes how news media select and emphasize aspects of reality while marginalizing other accounts in order to present a certain interpretation or explanation of events (Entman, 1993). If we think of newspaper front pages or home webpages, a single image is often selected to exemplify the news story and serve as a shorthand emblem of the news story. Traditionally, then, news photographs epitomize the *selectivity* of news framing. We discuss news frames and **visual framing analysis** further in Chapter 8 in our analysis of international conflict.

While the role of visuals in the construction of news frames has long been noted (Gitlin, 1980), and indeed their potential power over the textual content also acknowledged in media effects studies (Arpan et al., 2006), only more recently have systematic 'visual framing analysis' studies of protest prioritized visual information (usually photographs), in order to categorize and assess the varying emphasis on violence, destruction, antagonism or alternatively peaceful protest across a range of media. For example, see Perlmutter and Wagner's (2004) study of a 'news icon' from the anti-globalization protests in Genoa, or Batziou's (2015) content analysis of photojournalistic coverage of the 2008 Greek protests.

Self-made media by activists to raise awareness and mobilize support

How might activists spread their message and mobilize support beyond the 'immediate' surroundings, and bypass the gatekeepers of mainstream media? It is difficult not to think of the varied digital tools and platforms available to campaigners and activists since the development of the internet. In addition to the films, magazines, pamphlets, artwork and music that have long been part of activists' repertoire, the qualities of digital internet-enabled tools and platforms (accessible design, relative low-cost and instantaneous spread) have undeniably changed the way in which many activists plan their campaigns, especially if they are hoping to achieve transnational reach (Cottle and Lester, 2011). And with political uprisings from 2009 onwards labelled as 'Twitter' or 'Facebook' revolutions, the platforms' distinctive uses in mobilizing affinities, if not collective action, have been both celebrated and critiqued (see Papacharissi and de Fatima Oliveira, 2012; Wolfsfeld et al., 2013). But with our current focus on the visual, it is worth noting that only a few of these studies concentrate their analysis on the actual *content* of images shared (Olesen, 2013; Gerbaudo, 2015; Kharroub and Bas, 2016).

The sheer amount of information that can be uploaded and shared online at little cost means that such resources are not only suitable for instant awareness-raising and mobilization of support, they also form a counter-cultural (audio-visual) archive for future actions and generations. Writing on the Occupy movement, Thorson et al. (2013: 440) emphasize the varied roles for multiple platforms and the continuing value that such materials hold: 'Whether Occupiers regarded YouTube as a platform for publicity, sociality, circulation, or simply as a personal archive, the materials they uploaded contributed to a stock of resources available to publics associated with the Occupy movement.' The more direct communication enabled by such platforms might, however, also be ephemeral in nature – either deliberately so, or because such resources are still mediated through commercial companies and subsequent access for interested publics is not necessarily guaranteed.

Whether YouTube videos, Twitter handles and hashtags, Facebook pages, blogs, petitions, emails, memes or WhatsApp messages, each technology carries some of the baggage of its commercial creator but also the creative innovation, and sometimes critique, of the people who use such platforms. Tina Askanius (2010: 342) writes in relation to her work on video activism that such activities are indeed intertwined with technologies from their very conception: 'Activities are thus not simply locally bounded and then networked globally, but from the very outset a product of mobile transnational or translocal geographies of resistance and solidarity.' Notions of visibility and visuality are therefore central to this process of online activism.

Actions of resistance and solidarity are performed in part through activities such as uploading and sharing photographs, videos, mash-up and memes (Shifman, 2014), but it is also important to recognize how the intended meanings, symbols and affective responses are likely to change in different contexts and will not necessarily easily migrate or unify people in a wholly universal fashion.

And mixing it all up again!

It is useful, then, to separate the three forms of 'media' described above (visual activist materials; mainstream media portrayal; self-mediation activities) in order to understand the varying characters and styles of visuality and visibility with regard to protest and dissent. However, in reality such distinctions are messy and the categories of media do, of course, intersect and shape each other. The dynamics between protester-produced media, the mainstream media and social media are also now attracting scholarly attention.

For example, Mohamed Nanabhay and Roxane Farmanfarmaian's (2011: 573) study on the Egyptian uprising highlights the 'inter-related spaces of the **physical** (protests), the **analogue** (satellite television and other mainstream media) and the **digital** (internet and social media)' (added emphasis). In their analysis of YouTube videos uploaded and shared over the 18 days of the Egyptian uprising in 2011, starting with the mobilization of demonstrators on 25 January, Nanabhay and Farmanfarmaian write of an '**amplified public sphere**' created through this complex interplay, noting how the 'spectacle' phase initiated by the activists becomes a 'spectacular' phase once mainstream media broadcast such material to 'the nation and the rest of the world' (2011: 579). The interplay and intertextuality highlighted here also speaks to the image and the 'spectacle' as a productive space for creativity and mobilization, rather than assuming a passive audience. As DeLuca and Peeples argue in their theory of public screens cited earlier, the spectacles generated here come from the people rather than the rulers, with the aim of holding governments and corporations to account: 'Critique through spectacle, not critique versus spectacle' (2002: 134).

The overview above introduces the visual nature of protest as a vital feature in considering how activists communicate their identity, build solidarity across borders and mobilize support, and how they attract the attention of mainstream media. We have argued that it is not just the fact of dissemination or sharing that requires attention, but also the style or aesthetic characteristics of the visual and textual content. In analysing different spaces of mediated visibility, we show how activist images are employed to critique societal norms and build supportive publics. In each case, we understand the visual content as interacting with other communicative modes in each media context – for example, the message text in Twitter posts. So while we prioritize aspects of visual communication, we do not seek to isolate the images from the words in an overly artificial manner.

We now move on to our two case studies to show how aspects of visuality and visibility are mobilized in relation to two very different protest movements. Our first case study considers the visual tactics of the Russian punk collective Pussy Riot. Through a social semiotic analysis, we ask: 'What are some of the more properly aesthetic and stylistic traits found in such visual activism, and how are the semiotic resources of colour and texture, in particular, deployed in the immediate and mediated spaces where Pussy Riot perform their political actions?'

Our second case study looks at one of the many striking images from the Black Lives Matter protests, focusing more directly on the ways in which 'networked publics' participate in the early constitution of a *visual icon* on the social media platform of Twitter (Mortensen and Trenz, 2016; Boudana et al., 2017).

CASE STUDY 1: THE SOCIAL SEMIOTICS OF PUSSY RIOT

Our first case study in this chapter considers the 'image politics' of Pussy Riot across different communicative spaces. In tracing the different communicative contexts in which the performative and aesthetic dimensions of Pussy Riot operate and move, our intention is to explore how visual politics play a key role, and how the meaning and ethos of Pussy Riot are transformed and reconfigured in their movements across media genres and platforms. We offer here a brief social semiotic analysis, with *colour* and *texture* as the highlighted semiotic resources, alongside an interest in the embedded or situated nature of visibility and visuality offered across different media platforms and actions on the ground.

Research questions and approach

For this case study, we follow a social semiotic approach and so the emphasis here is on mapping the meaning-making potentials of various semiotic resources across different media (see Chapter 2 for further details). Specifically, we explore the semiotic practices of Pussy Riot to better understand how they articulate their identities, demands and desires through intertextual references to existing signifiers (of punk aesthetics, for example) and how their knowing, and even satirical, utilization of striking visual communication draws attention to their campaigns. Our two key texts for methodological inspiration here are Gunther Kress and Theo van Leeuwen's (2002) article on **colour** as a semiotic resource, and Emilia Djonov and van Leeuwen's (2011) article on the semiotic potential of tactile and visual **texture**. Therefore, we ask the following questions:

- How do colour and texture work as semiotic resources to signify the punk ethos of Pussy Riot?
- How are the political actions of Pussy Riot performed in the spaces of the immediate, mediated and self-mediated?

So who are Pussy Riot?

Pussy Riot formed in 2011, ostensibly in response to Vladimir Putin announcing that he would run for president once again. While primarily concerned with the concentration of power under Putin and the corruption of democratic values in Russia, the influences of Western feminisms and Western artists are manifest in the visual and musical styles of Pussy Riot. With their political actions intersecting with performance art tactics and other avant-garde arts activism, Pussy Riot were arguably peripheral actors in the dissident scene in Russia until the arrest and subsequent imprisonment of three members, which ironically brought them the international media attention they had hoped to provoke. When Nadezhda (Nadya) Tolokonnikova, Maria (Masha) Alyokhina, and Yekaterina (Katya) Samutsevich were 'unmasked' and put on trial for hooliganism in 2012, the authorities appeared to be aiding in the latest media spectacle they had devised. Nadya's husband, Pyotr, provides an apt metaphor: 'Pussy Riot is a *machine placed inside the media* [. . .]. It's designed to draw attention in the West to what's going on in Putin's Russia. And the group has succeeded' (cited in Bennetts, 2014: 147, emphasis added).

Let's step back slightly, then.

Pussy Riot had already performed a number of 'actions' when they decided to provoke the powerful institutions of Government and Church with their 'punk prayer' at the Cathedral of Christ the Saviour in Moscow on 21 February 2012. This was around their fourth 'action', following disruptive performances in luxury boutiques or in support of arrested protesters outside detention centres;

and, crucially, this time they had invited trusted journalists along to help spread the word. With the intention of critiquing the way in which Putin had used the cathedral as a backdrop for his own political photo opportunities, Pussy Riot were deliberately placing themselves in the sacred space of the altar where women are not usually allowed to enter. In Katya's closing statement at the trial, she questioned why Putin had chosen to exploit the Orthodox religion and its aesthetic, using the cathedral as a 'flashy backdrop' – it was this political image, with the Patriarch supporting Putin in highly orchestrated events, that Pussy Riot hoped to confront and reveal as false. This was an attack on symbols and a retort to the particular political uses of aesthetics (Figure 7.1).

FIGURE 7.1   Pussy Riot perform their punk prayer in the Cathedral of Christ the Saviour in Moscow on 21 February 2012. Copyright: Sergey Ponomarev/AP/Shutterstock. Reproduced with permission.

Findings: becoming Pussy Riot across immediate and mediated spaces

The clashing bright colours of the balaclavas, tights and dresses disrupted the symbolism of the splendour inside the cathedral. In this way, the gold, marble and velvet are purposefully juxtaposed with their own shambolic and even degenerate style. Once captured on video by their own group and photographed by reporters, they soon uploaded the images to their YouTube channel. The embodied and local performance fulfilled the *immediate* objective of invading such sacred space, but their editing and sharing of the video online was also crucial in mobilizing support and reaching out to a networked global audience.

Colour as a semiotic resource

In their exploration of colour as a communicational resource, Kress and van Leeuwen (2002) set out a number of features that help us to think about a 'grammar of colour' as part of the signifying practices we might choose to employ in making meaning. As they write, colour cannot exist by itself but combines with other modes (such as texture, as we'll explore below), to provide certain associations (including those with highly symbolic and emotive values) and distinctive qualities (light/dark, saturated/desaturated) which serve as meaning potentials. If we think of the conventions of interior or fashion designers, their expertise is in creating colour harmony and positive associations, but for activists colour can serve other communicative and creative purposes.

Marian Sawer (2007: 40) has written eloquently on the importance of political colours for activists and social movements, with her study focusing on the women's movement: 'The colours not only served the purpose of visual identification with "the cause" and the outward display of values but also played an important role in sustaining a sense of community.' Sawer stresses the emotional identification with such visual symbols and how they play a part in the contestation of public memory, but also touches on the relative paucity of studies about the importance of political colours: 'considering the highly visual nature of these strategies, the neglect of the role of political colours and related symbols is surprising' (2007: 54).

The colours of the Pussy Riot masks or balaclavas are crucial in this case. The balaclava stands as a visual metonym for Pussy Riot, one that is both flexible and resilient: it is easily adopted by others, but is also instantly identifiable and distinctive. Unlike the black masks of anarchists or terrorist groups, the brightly coloured masks signal dissent while being unthreatening; as we noted above with reference to the work of Pollyanna Ruiz (2013), the mask connotes solidarity and collectivity through its anonymity. In an interview with author Masha Gessen, members of Pussy Riot expressed this very idea:

> "Being Pussy Riot is like being Batman," one participant told me. "*You put on the mask – and you become Pussy Riot*. You take it off – and you are no longer Pussy Riot." The mask, in this case, was a balaclava. It could be any colour as long as it was bright and one wore tights to not match it. (Gessen, 2014; emphasis added)

We can even read the mask as a form of media itself here. In his work on the Clandestine Insurgent Rebel Clown Army, who specialize in creating bodily engaged, emotive and disruptive performances at meetings such as the G8 summit in Scotland in 2005, Paul Routledge writes of the clown face as 'a form of tactical media' (2012: 436). The masks give anonymity and protection from the police, but they also offer a satirical response to social masks of convention. As Routledge claims, there are multiple functions at play here – freeing up personal inhibitions, exaggerating absurdity, undermining intimidation, magnifiying commonality and affinity, and finally engaging the media.

We can make similar claims for the Pussy Riot masks – there is the performative act of putting on the mask and becoming Pussy Riot. The saturated clashing colours help the Pussy Riot masks and 'uniform' to become a visual symbol and even an icon – one that seems ready-made for the global media. As noted, colour cannot exist on its own, and here we point to texture as a complementary semiotic resource deployed by Pussy Riot.

The semiotic potential of texture

In their subversive, ironic actions and self-mediating practices, Pussy Riot intend to layer meanings through various semiotic resources – appropriating or playing with the recognized visual tropes of punk, DIY aesthetics, disharmony and feminist art (see Smith-Prei and Stehle, 2016). So while they are operating in a certain moment in Russian political history, they are also borrowing from recognizable cues with universal cross-media power. The textures of their masks and shiny guitars are also potentially meaningful in the semiotic landscape. As Djonov and van Leeuwen (2011: 541) write, the quality of texture, even within mediated images, 'suggests an illusion of tangibility', which can be achieved 'across different media and can have tactile as well as visual and aural manifestations'. As suggested with the meaning potentials of colour, texture can 'function textually/ cohesively to establish harmony or contrast between elements', and can signify, for example, 'comfort, warmth and elegance' (2011: 545, 547) in its qualities of smoothness, roughness, dryness, or stickiness.

FIGURE 7.2 Image of protester showing solidarity with Pussy Riot, here with the additional connotations of the leftist movements of Central America. Posted on Jeanette Martin's (@Shutterjet) Twitter page (https://twitter.com/Shutterjet/status/236496732186353666). Reprinted with kind permission.

The simplicity of the cultural icon of the woolly colourful balaclava means that solidarity can be easily expressed through wearing one and by sharing the images on social media (see Figure 7.2).

Pussy Riot appeared to be tapping into an ethos of playfulness and craft that had assumed an ideological global resonance. As Susan Luckman has argued (2013), there has been a renewed popularity of the handmade and of craft production, with the former disparagement of often women's craft practices now replaced with the wide acceptance of handmade objects as aesthetically desirable, especially among younger consumers. It could be that part of Pussy Riot's Western appeal accidentally chimed with this renaissance of the handmade and its associated appeals to empowerment and authenticity. The meaning potential here comes from the ethos of handmade crafts and vintage clothing *as set against* the cultural homogeneity of shop-bought items (and their reliance on cheap labour); instead accentuating the sensorial quality of the materials and the values of re-fashioning or recycling items.

Earlier we cited Pussy Riot members who stated, 'You put on the mask – and you become Pussy Riot'. But is it possible to become Pussy Riot by *buying* a mask, rather than recycling an old ski mask? Is it inevitable that some entrepreneurial person sees a market in selling Pussy Riot-themed balaclavas and so takes to Etsy? For US$30 you too can own a Pussy Riot inspired balaclava from various sellers on the Etsy website.

The fledgling Women's March movement that greeted the inauguration of Donald Trump in January 2017 further demonstrated how a visual symbol offers a way to show solidarity and support across the world. The 'pussyhats' worn by protesters were distinctively crafty, imperfect and colourful (usually pink). According to the 'Pussyhat Project' website, the hat was conceived to make a 'powerful visual statement' through the collective craft practices associated with women. More community-driven and certainly less combative in tone when compared to Pussy Riot, the Pussyhat Project aims to build a global movement and, at the time of writing, hopes the symbol of the pussyhat remains widely visible at International Women's Day marches.

Case study conclusion: from punk activism to global brand

Pussy Riot perhaps do not easily fit into the social movement paradigm, operating within the realms of art activism, but also situating themselves within a global movement of feminists and anarchists. They provide an expressive form of solidarity for outsiders, challenging the hegemony of an increasingly authoritarian state, but through globally mediated technologies and with *Western* audiences and citizens in mind.

We have explored here a particular historic moment, in which the cultural icon of the colourful balaclava worked as a semiotic resource alongside various activist-related actions, mediations and playfulness, or even deliberate absurdism.

But all protest practices adapt and repertoires need to evolve according to changing circumstances and opportunities. The 'unmasking' of Pussy Riot has led to a new era in their campaigning where new visual tactics are required. In a way, Pussy Riot were trickster, clowning, jester-like figures, but there was an associated danger in performing their own triviality or tormented outsiderness. Russia has its own particular tradition of 'holy fools', as Marc Bennetts describes them, 'half-witted prophets given licence in medieval times to say out loud what others could not' (2014: 141). In her closing statement at the trial, Nadya referred to this holy foolishness, but also linked it to their punk ethos: 'We were searching for real sincerity and simplicity, and we found these qualities in the yurodstvo [the holy foolishness] of punk' (cited in Bennetts, 2014: 141). Pussy Riot captured the attention of the global media through their colourful (un)holy foolishness, but whether their 'brand' of activism will have any longer-term effects remains to be seen.

What we can note is how Pussy Riot's visually creative political actions worked across various spaces: the *immediate* political performances on the ground, where their female punk presence was designed to provoke authorities; and in the **self-mediated** spaces of their own YouTube channel and in the **global media** uptake of both their style and their cause.

ACTIVITY

> The *V for Vendetta* mask has become an international symbol of protest and resistance. Writer Alan Moore and artist David Lloyd created the *V for Vendetta* graphic novel series in 1982, with a film version released in 2006. Associated most closely with the hackers' group Anonymous, the mask is also widely used at other protests. How might you design a semiotics-based study on this mask, drawing on the approach above?
>
>> The Guy Fawkes mask has now become a common brand and a convenient placard to use in protest against tyranny – and I'm happy with people using it, it seems quite unique, an icon of popular culture being used this way. (David Lloyd, graphic novel artist, cited in Waites, 2011)

CASE STUDY 2: EXPRESSIONS OF INJUSTICE AND VISUAL ICONS: #BLACKLIVESMATTER

As mentioned above, our second case study examines one of the many striking images from the Black Lives Matter protests, focusing more directly on the role of

Twitter as a space in which a 'public discourse of images' (DeLuca and Peeples, 2002: 133) is performed through the collective circulation, discussion and repetition of tweets that include a certain photograph. In these practices, tweeters co-create a 'visual icon' of protest that is believed to communicate something powerfully symbolic.

A brief history of the Black Lives Matter movement

The Black Lives Matter movement first formed and gained prominence after the acquittal of George Zimmerman for the shooting of Trayvon Martin, a 17-year-old African-American teenager shot dead by the neighbourhood watch volunteer after he popped out for sweets and a drink from a local store. While Martin's death in February 2012 prompted debates on racial profiling, laws on citizen safety (known as 'Stand Your Ground law' in Florida) and on police reluctance to charge Zimmerman, it was the trial and acquittal of Zimmerman in July 2013 that prompted the hashtag campaign #BlackLivesMatter, co-founded by activists Alicia Garza, Patrisse Cullors and Opal Tometi.

The impetus for the hashtag campaign to move beyond social media platforms and build an identifiable style working with artists and designers came with the police shooting of Michael Brown on 9 August 2014 in Ferguson, Missouri. The sense of injustice grew in strength as further cases across the US appeared to demonstrate a police readiness to injure and kill (young) African-American men who did not appear to pose any immediate threat. Even before Michael Brown's violent death in Ferguson, mobile phone footage had come to light which depicted a number of New York City policemen holding down Eric Garner, with one officer, Daniel Pantaleo, using a 'chokehold' that restricted Eric's breathing. Eric Garner died on 17 July 2014 after repeatedly telling the officers 'I can't breathe' while being held down on the pavement. This case really came to international attention in December 2014 when a grand jury decided not to indict Pantaleo, sparking further protests, including 'die-ins' where protesters lay across roads in various US cities, as well as in London.

The fact that his friend Ramsey Orta had captured the attempted arrest of Eric Garner on his mobile phone meant that a global audience could bear witness to Garner's struggle against the police. First made public by the *New York Daily News*, the footage circulated on various platforms, through peer-to-peer networks, YouTube and mainstream news websites.

Research questions and approach

The nature of 'iconic' photographs is explored in more detail in Chapter 8, but our interest here are the ways in which 'networked publics' participate in the early constitution of 'visual icons' in a digital media ecology where media and political elites might be following rather than leading the commentary (Boudana et al., 2017; Mortensen and Trenz, 2016). As Mette Mortensen and Hans-Jörg Trenz (2016: 349) argue, citizens and activists now play a part in mobilizing and framing certain images as 'iconic': 'By promoting certain images, these non-elite actors contribute to framing how the public addresses and imagines important tensions and challenges in society.' In this case study, we ask:

- How are the political significance of the image and its 'iconicity' discussed and mobilized on Twitter?
- Is the photo altered or repurposed in shared tweets and what does this tell us about its significance (political, cultural, aesthetic, affective)?

In the limited space available here, we do not develop a full-scale quantitative social media analysis of the Black Lives Matter campaign, although others have contributed studies of this nature (see Freelon et al., 2016; Jackson and Foucault Welles, 2016). Instead, we capture a smaller subset of tweets inspired by a particular photograph which was soon hailed as 'iconic' and morphed into cartoons and other illustrations within days of its publication and circulation. Rather than taking an amateur or citizen image as our focus here, we are referring to a photograph by Jonathan Bachman, whose photograph of Ieshia Evans's arrest 'went viral' after he had filed it with Reuters from a parking lot using a hotspot from his phone (see Bachman's interview in Jones and Frizzell, 2016).

There are a number of challenges to collecting tweets for visual research, as search tools usually identify linguistic information (e.g. hashtags or usernames) rather than the photograph or image itself. We used the tool Mecodify (www.mecodem.eu/mecodify/) to select tweets using the hashtags #BlackLivesMatter, #BLM and #BatonRouge. To capture the later public discussion, we also included 'Ieshia Evans' and 'Bachman', along with other spellings which also circulated ('Iesha', 'Bachmann').

We acknowledge that we might not have captured every single tweet about the photograph, but this method certainly enabled us to track the related activity and discursive nature of the responses of those who shared the photograph in the context of the Black Lives Matter campaign. Indeed, by choosing to include the hashtags or names alongside the images or links, tweeters are signalling that they intend to play a part in a public conversation, or following Deluca and Peeples, to participate in a 'public discourse of images' (2002: 133).

Findings: a visual icon for #BlackLivesMatter

Visual materials are crucial to the campaign, especially in the evidential quality of amateur photographs and video. The capture of the police shootings and brutality on mobile phones and the sharing of mainly video content have undoubtedly helped to raise awareness. Such images spur those compelled by outrage and anguish on to the streets, in many cases proclaiming to have never before taken such politically motivated action. Here again, the images of protests demonstrate the desire to be publicly visible, as an expression of solidarity and to provoke change in the justice system.

Acknowledging how the news media outlets also consume and remediate the images of their protests, activists and supporters choose colourful clothes, symbols of peace, repetitive 'hashtagged' slogans and the names of those killed written on banners, in an effort to form a collective visual memory. In their report on how Black Lives Matter ignited a national conversation on police killings of unarmed black citizens, Deen Freelon and co-authors point out the very nature of police brutality as an issue well-suited to an internet-based campaign: 'Unlike wealth or income inequality, police brutality is concrete, discrete in its manifestations, and above all, visual' (Freelon et al., 2016: 82).

The photo icon we have chosen to discuss in this case study circulated in social and mainstream media following the shooting of Alton Sterling in Baton Rouge on 5 July 2016. Sterling's death was captured by a number of bystander cameras and CCTV as he was shot multiple times by police officers who held him on the ground. In response to this shooting and the killing of Philando Castile the very next day (the immediate aftermath of which was livestreamed on Facebook by his traumatized girlfriend, Diamond Reynolds), this period marked an increase in tension and new protests in both Baton Rouge and Dallas.

'What a powerful photo'

FIGURE 7.3   The original photograph by Jonathan Bachman (Reuters) of Ieshia Evans at Baton Rouge, posted on Twitter initially without crediting the photographer and with the caption 'What a powerful photo'. Copyright: Jonathan Bachman/Reuters Pictures. Reproduced with permission.

So what is it about this photo that captured the imagination of those who shared it online?

The off-centre balance in the composition casts two heavily protected police officers, who seem to be moving backwards, against the calm dignity of Ieshia Evans with her summer dress blowing in the breeze (Figure 7.3). Holding her phone in one hand, she offers the other hand to the arresting officers whose movements appear clumsy and unstable in their heavy body armour. A crack in the road splits the protagonists, as if perfectly planned by the photographer to symbolize the sheer force of Evans's poise, which might just break the tarmac if we were to accept the suggestive supernatural allure of the composition. Indeed, the 'superpower' resonances of the image were soon debated on Twitter and across the media, as we note below.

We trace the photograph's appearance on Twitter back to @BonsaiSky, who simply captions the image 'What a powerful photo' on 10 July 2016 at 8.15am (https://twitter.com/BonsaiSky/status/752159435808702464/photo/1). This starts a collective effort by other users to track down first the photographer and then the name of the woman pictured. Within half an hour, user @RobertLoerzel finds the image credit for Jonathan Bachman through a Google search, and expresses surprise at Reuters' apparent lack of awareness that the image 'has gone viral': 'Everyone is talking about this great photo, but @reuterspictures and the photographer seem to have been silent about it on Twitter' (https://twitter.com/robertloerzel/status/752166868786434048).

So how is the image's 'iconicity' mobilized on Twitter?

In addition to simply sharing the photograph, with captions emphasizing the 'power' of both the photograph itself and of Ieshia Evans's grace, composure and bravery in contrast to the police officers, Twitter users start to suggest linkages to other critical moments, rights campaigns and national symbols. In doing so, the public and journalists collectively start to shape a shared understanding of the significance of image in relation to historical events and cultural mythologies. For example, one user, @G33KDad, posts a mash-up collage of the photograph with Martin Luther King Jr appearing in the far right of the frame behind Ieshia, encased in a shimmering light and providing a raised hand that appears to repel the officers (Figure 7.4). The spectral figure suggests a spiritual affiliation with the earlier hero of the civil rights movement while also alluding to the superpower associations that Evans appears to possess. The tweeter name checks the photographer and also later includes Ieshia Evans's name, thus following certain codes of politeness alongside the codes of association indicated by the #BlackLivesMatter hashtag and three fists icon.

Soon, there are other examples of collages, mash-ups and illustrations. The rapidity of the creativity demonstrates the compression of time between the production of the image and its reappropriations and refashioning into

FIGURE 7.4 @G33KDad's photo-collage of Ieshia Evans protected by the ghost of Martin Luther King (https://twitter.com/G33KDad/status/752373501290438656).

other forms. Another posting that captures the oft-mentioned linkages to other moments of peaceful resistance comes from @Matt_Twomey, aligning the photograph not with a single critical moment, but with five other iconic images, captioned: 'When overwhelming force is met with fearless grace and dignity' (Figure 7.5). Linkages of this nature were also promoted in the mainstream media – for example, a *Foreign Policy* article by David Rothkopf which merged the Bachman photo with the 1989 picture of the 'tank man' of Tiananmen Square (see bottom left image in Figure 7.5). It is worth noting that such comparisons also received criticism from other social media users, often concerned about how a visual union of two disparate events obfuscates, rather than enhances, the political meaning and significance of either image.

The final tweet we include in our very selective overview consists of a cartoon by Andy Marlette (@AndyMarlette) who adapts the original photograph so that Uncle Sam now bows down before Evans, offering what looks like a peace lily. Figure 7.6 depicts the later retweet by @hellocojo who pointedly adds '#sayhername iesha evans' to Marlette's original caption of 'The girl in Baton Rouge'. *Naming* here becomes part of the public recognition of her political agency and also references a #BLM related advocacy campaign called #SayHerName, working to combat state violence against black women.

The cartoon enables a reimagining of the captured photographic moment where Ieshia stands defiantly in the light, as a sketched darkness falls around Uncle Sam, as if he is attempting to hold back the gloom across his shoulders and back. His necktie is loose and his gaze is firmly towards the ground, suggestive of a respectful yet weary surrender to a new national icon. It is doubtful that conservative white America is willing to bow down to Ieshia as depicted, but the figurative style of the cartoon allows for an imagined and symbolic vision of a peaceful way forward.

Table 7.1 below briefly summarizes the visual and discursive techniques we observed that work to mobilize the image as 'iconic'. In following the patterns of circulation, repetition, refashioning and reframing in this way, this analysis tracks the early interest and impetus created by

FIGURE 7.5 Twitter user @Matt_Twomey's visual associations with past resistance movements (https://twitter.com/Matt_Twomey/status/752568225402060801).

FIGURE 7.6 Twitter user @hellocojo adds '#sayhername iesha evans' to the cartoon: https://twitter.com/hellocojo/status/752855302081568768 (original posting by Andy Marlette: https://twitter.com/AndyMarlette/status/752557568980054020/photo/1).

'networked publics' as they negotiate the aesthetic, cultural and political power of the image in the digital media environment.

TABLE 7.1 Mapping the mobilization of the Ieshia Evans photograph as a visual icon.

Values attributed to the photo and Ieshia Evans	Intertextual linkages and political significance	Frequently used media links
Powerful	Martin Luther King's hand (historical civil rights movement, campaign for injustice); likened to Rosa Parks not giving up her seat	Jonathan Jones's piece in *The Guardian*
Instant classic		Evans's interview with Gayle King on CBS and with 'Young Minds Can' organization
Extraordinary		
Making a stand		
Critical moment	Linked to 'tank man' of Tiananmen Square verbally and visually (e.g. in *Foreign Policy* article)	*Foreign Policy* article on US losing moral high ground
Legendary		
Greatest photo of 2016		Evans writes an op-ed for *The Guardian* (most retweets)
Graceful		
Defiance	'modern-day Statue of Liberty' (*Washington Post*)	Report that Baton Rouge prosecutor has dropped the charges against the 102 people arrested
Strong	Cartoon with Uncle Sam kneeling before her	
Awesome		
Superhero		Bachman's interviews with *Time* magazine and *The Guardian*
Bravery		
Inspiration		

Constructing an iconic image

In February 2017, Bachman's photograph of Ieshia Evans was awarded first prize in the Contemporary Issues, single image section of the 2017 World Press Photo competition, the kind of recognition that further imbues the image with iconic status. In the pre-internet age, such status was conferred on certain images as they repeatedly circulated in newspapers, magazines, television shows, reviews of the year, photography books and art galleries. This is not to suggest a linear journey, but each medium and genre arguably augments a further permanence and prestige value, taking a position in the archives that construct shared cultural memory.

Public conversations among citizens were always a significant part of this photographic journey, but now such contributions and patterns of dissemination are recorded on social media platforms, and, in temporal terms, citizens might even lead the way in co-constructing the value, meaning and 'iconic power' of such images in public life. Additionally, such exchanges are enhanced and amplified through global media and personal networks, so that the signification and significance of the image are constituted, mobilized and reframed within ever more diverse public spaces.

Crucially, and perhaps in contrast to earlier photographs deemed to be 'iconic', within a very short space of time Ieshia Evans had not only given interviews and posted her reactions on Facebook, she had also written an op-ed piece for *The Guardian* (Evans, 2016). We cannot say for sure if the publicity around the image played a part in the Baton Rouge prosecutor dropping all charges for those arrested, but we can certainly say that the visually striking nature of the image led to Ieshia Evans's voice being heard. This was not a job she 'applied for', to use her own words in the CBS interview, but she has become a vocal campaigner for the movement.

Case study conclusion: the right visual icon for Black Lives Matter?

It is possible that the image resonates precisely because it counters the violent and fear-mongering pictures that had become the mainstay of mainstream reporting about the Black Lives Matter movement. There is no violent struggle for life pictured here; indeed, there almost appears to be a force-field around Ieshia as the heavily armed and protected officers attempt to arrest her. However, some have questioned whether this captured moment of calm and even beautiful dignity disavows the sense of outrage and injustice at black men's deaths that we should be responding to in more politically engaged ways: 'Yet is this picture popular because it expresses the truth, or because it softens it?' (Jones and Frizzell, 2016).

The gender politics of both the image and the subsequent public discussion are other possible avenues for further exploration that we could not cover here. Ieshia Evans has become an accidental icon of the Black Lives Matter movement, offering a *picture* of graceful dignity and a *voice* of peaceful resistance at a crucial moment.

CHAPTER CONCLUSION: PROTEST IMAGES AS TOOLS FOR POLITICAL ACTION

In this chapter we have explored just some of the ways in which visuals are deployed by activists and social movements to stir emotions and mobilize support. The particular communicative capacities of visual materials and mediated images help activists to build recognizable identities across borders and to encourage a shared imagining of a different way of being. Of course, these activities are not always progressive in nature and it is important to explore how symbols and flags can also become rallying features for violent and nationalistic movements. In sum we hope that readers will take away the following:

- In analysing the aesthetics of activism in three interrelated spaces (immediate, mediated and self-mediated) we are able to trace how the meanings of protest visuality and visibility are conveyed, interpreted and transformed.

- We can start to understand the significance of visual materials and images through examining their use as semiotic resources.
- Digital technologies allow us to track how (accidental) protest visual icons are co-created, shared and discussed via 'public screens', and might even engender 'networked publics' who share common fears and hopes.

POTENTIAL FURTHER RESEARCH

This chapter has focused its attention on the mobility of protest images across varied media, offering templates and ideas for possible future research. But another way to explore the significance of imagery for activists or protesters would be to conduct research with those who make or share such images. Guiding questions could be: How do activists describe and value their own visually oriented practices? Is the style or aesthetics of a campaign central to their strategies and organization?

FURTHER READING

A number of special issues provide varied essays on aspects of protest and visual images. These include one entitled 'Advances in the Visual Analysis of Social Movements' for *Research in Social Movements, Conflicts and Change* (Doerr et al., 2013) and 'Picturing Protest: Visuality, Visibility and the Public Sphere' for *Visual Communication* (Rovisco and Veneti, 2017).

8 PICTURING INTERNATIONAL CONFLICT AND WAR

The idea of media images as 'weapons' of war has become a familiar refrain, with wars and conflicts fought out in the 'mediatized' battle space as much as on the battlefield (Parry, 2010b). Sophisticated and well-funded militaries now find that they engage in a 'battle for hearts and minds' over the internet, with rebel armies, insurgent groups and terrorists appealing to supporters through 'homemade' media with global reach. This chapter explores how war images hold the potential to convey some of the most devastating consequences for humans living through violent conflicts, but also how such images are filtered through various mediating processes, whether journalistic practices and conventions, or via the personalized spaces of social media platforms.

This chapter:

- summarizes the role of images in portraying war;
- explains why images of conflict offer such potent 'visual icons';
- conducts a visual framing analysis of ISIS propaganda in the UK press (Case study 1);
- analyses survey responses to an iconic image from the Syrian war: Omran Daqneesh, the boy in the ambulance (Case study 2).

As this chapter is concerned with the particular communicative capacities and qualities of news photographs, we explore the fascination with their journalistic value as both objective record and ideological tool, and how such paradoxical attributes have been debated by photography scholars over the years. Images depicting the dire human consequences of war are often those that garner prestigious awards, and also become emblematic or 'iconic' as they persist in media commemorations of significant historical events. Before turning to our two case studies we summarize the useful approach of **framing analysis** (which helps us to analyse how events are constructed and interpreted by newsmakers, both visually and verbally), and discuss how certain images become *visual icons* in both traditional media but also through citizen sharing on social media platforms.

THE ROLE OF PHOTOJOURNALISTIC IMAGES DURING WARTIME

The vital role of the news media during times of war means that it is necessary to study how conflict imagery provides complex layers of meaning, with the potential to reinforce or question the dominant understanding of a war or conflict. Past research has pointed to various concerns about the role of media imagery during wartime, with news visuals in particular cast as 'weapons' or 'tools', engaged strategically by a number of political and military actors. For instance, visual culture scholar Nicholas Mirzoeff notes in *Watching Babylon* (2005: 73) that the war image became a 'smart weapon' during the 2003 invasion of Iraq: 'Consequently, the images of the war were not indiscriminate explosions of visuality but rather carefully and precisely targeted tools.'

Reiterating Mirzoeff's observations, Barbie Zelizer asserts that 'journalism's images of war provide a strategically narrowed way of visualising the battlefield' (2004: 130), and she sees this as particularly problematic when images are selected due to their symbolic power. Rather than simply adding to the information on the page, the symbolic quality of the photograph serves to couch what is being depicted in broader, generalizable frames. The point here is that news photographs function as documentary evidence in their 'objective' recording of events, but also perform symbolic and political functions, especially where they resonate with pre-existing cultural myths (often related to notions of patriotism and national superiority).

This **connotative** power is viewed as crucial by Zelizer, with familiar imagery recycled, either literally from photographic archives or compositionally through recognizable themes and portrayals. Here, the informational role of the image is superseded by its symbolic force. Pictures are chosen based on aesthetic appeal and therefore reinforce conceptions of the 'good' war image through repetition and conventionalized depictions, whether they are of fetishized weaponry or heroic soldiers. Such concerns about emblematic and familiar photographs particularly come to the fore in those images widely labelled as 'iconic'; and we discuss the role of 'visual icons' of war in more detail later in the chapter.

An objective record of the pain of others?

In addition to the critical attention paid to the war image as a strategic tool of the military, designed to glamorize heroism and military prowess, scholars have long expressed concerns about the ideological function served by images of human suffering and the specific capacities of photography when it comes to promoting understanding. In her short book *Regarding the Pain of Others* (2003), cultural critic Susan Sontag revisits some of her earlier claims made in her famous collection of essays, *On Photography* (1979).

Sontag refers to the contradictory features of photography as an advantage: 'Their credentials of objectivity were inbuilt. Yet they always had, necessarily, a point of view' (2003: 23). Yet she also sees this dual role as both 'objective record' and 'personal testimony' as a 'sleight of hand' (2003: 23). This is an optical trick whereby our attention is diverted to the accepted evidential nature of the news photograph, often because the latent ideological message is so well known and so recognizable as to be barely noticeable.

In her later book, Sontag is primarily concerned with the role of harrowing images and their limitations for promoting understanding. In contrast to narratives, which can help us to understand, the photographs 'do something else: they haunt us' (2003: 80). However, Sontag is perhaps not so critical here as she may initially sound, as she believes that this role is crucial, despite the failings that photographs may have if we expect too much of them:

> Let the atrocious images haunt us. Even if they are only tokens, and cannot possibly encompass most of the reality to which they refer, they still perform a vital function. The images say: This is what human beings are capable of doing – may volunteer to do, enthusiastically, self-righteously. Don't forget. (2003: 102)

For Sontag, photographs have the 'deeper bite' in the role of remembering, especially in comparison to the televisual flow of images that 'keeps attention light, mobile, relatively indifferent to content', and she amends her earlier assertion that photographs of suffering can have a neutralizing effect: 'As much as they create sympathy, I wrote, photographs shrivel sympathy. Is this true? I thought it was when I wrote it. I'm not so sure now' (2003: 19, 94). Sontag's reflections and later change of heart demonstrate just how complex and unsettling our task is in trying to understand the effects of such images.

Sontag's writing on photography, along with other theorists such as John Tagg, John Berger and Allan Sekula, has been very influential in both taking photography seriously as an artistic and cultural form, and in developing a critique that is often imbued with misgivings and uncertainties, especially when it comes to picturing the human consequences of war or disasters. To summarize some of the points above, which recur throughout photography writing: photographs are both a trace of reality before the lens, but also highly selective in what the photographer chooses to show; they are both objective and deeply personal; they have a power to haunt us, but also require words to help us understand; they let us bear witness to what humans do to each other, but can also neutralize our sympathy. These are just some of the contradictory features that continue to fascinate and frustrate photography scholars.

And this critical account of photography comes to the fore when considering war photography. An emphasis on limitations, disregard and despair has traditionally dominated the academic literature when assessing efforts to achieve ethically responsible representation of suffering (encompassing issues such as voyeurism, othering, spectacle), with perceived constraints grounded in the pictures' 'anaesthetizing' or 'aestheticizing' qualities (Sontag, 2003; Chouliaraki, 2006). However, scholars such as Robert Hariman and John L. Lucaites (2007a, 2016) and Susie Linfield (2010) have countered some of the more pessimistic critiques of the capacity of photography to induce empathy and prod consciences: 'Today it is, quite simply, impossible to say, "I did not know": photographs have robbed us of the alibi of ignorance' (Linfield, 2010: 46). We cannot claim ignorance when photographs of suffering are so readily available to us, but it is impossible to predict if *seeing* such images will lead the spectator to empathy or action. However, we can examine the kinds of war images that are selected and 'framed' in the news.

Visual news frames

As part of their interpreting practices, newsmakers draw upon value-laden words, stereotypes, metaphors and symbols, either consciously or unconsciously, in ways that connect with recognizable templates or pre-existing ideas, and so offer ways to make sense of new and often ambiguous events. Embedded within such a news 'package' images offer a particular form of realism, emotional authority and cultural resonance. In the context of war, this struggle over meaning is crucial for the antagonists who are engaged in a battle for 'hearts and minds' in the global mediascape, promoting their own moral and military superiority and maintaining morale on the homefront and battlefront.

Photographs and news images play an important role in such 'news frames', both appearing to present actuality and performing symbolic and political functions at the same time (see Chapter 2 for an introduction to visual framing analysis). For Messaris and Abraham (2001: 220), such qualities suggest a central role for visual images in framing research:

> The iconic ability to seemingly reproduce nature means that visual images are capable of producing documentary evidence to support the commonsensical claims of ideology, and in turn to use the very appearance of nature (seemingly factual representations) to subtly camouflage the constructed, historical, and social roots of ideology.

We can see then that framing analysis is concerned with both the conscious and unconscious reinforcement of dominant ideologies through the selection or omission of value-laden elements or symbols. **Visual framing analysis** places the focus of research on the use of photographs or images as they are printed and textually embedded in news coverage. In this case, the persistent patterns of 'selection, emphasis and exclusion' (Gitlin, 1980: 7) of imagery rather than written text or spoken words serve as the *primary* focus of analysis. In sum, visual framing analysis examines the ways in which the news narrative is performed through the selections, repetitions and patterns in visual coverage, in order to gain insight into the often tacit and unstated cultural and political sensitivities of the newsmakers and their imagined readers.

> ## FOCUS
>
> Think about how you might design a content or framing analysis with the news photograph as your primary unit of analysis. What elements or measures might you want to include when analysing images within a news context?
>
> **Contextual factors**: headline, caption, other text, source for the image (news agency, archive).
>
> **Layout**: position on the page, colour, size, juxtaposition with other news stories and images but also adverts.
>
> **Compositional:**
>
> - Who is photographed (number, gender, age)?
> - Angle and distance – what insights into the personality are offered?
> - Do the subjects look directly at us (gaze)?
> - How does the photo work compositionally to draw our eyes to a particular feature or person?
> - Does the shot appear posed for the camera?
> - What kind of behaviour is depicted?
> - Does the photograph elicit an emotional reaction from us? If so, how?
> - Is graphic imagery included (bloodshed, death)?
>
> In some cases, the information coded for is 'manifest' or literal (headline text, distance), but the questions above also require a level of interpretation, asking for connotative judgements to be made. Even if a single person is conducting the analysis, it is worthwhile developing a set of instructions or coding conventions to ensure a consistent approach to the material. It is important to also keep in mind *why* you are gathering such aspects and how they fit with ideas about the images' meaning-making capacities. For more details on how make such connections, see Gunther Kress and Theo van Leeuwen's *Reading Images: The Grammar of Visual Design* (2006) or Gillian Rose's chapter on content analysis in *Visual Methodologies* (2016).

VISUAL ICONS

As we touched upon in Chapter 4 and in Chapter 7, 'iconic' images are those deemed to exemplify the values or other cultural, political or social significances of a particular moment in history, usually depicting human subjects as archetypal or symbolic figures. The iconic status of photographs is performative to some degree – traditionally, it is 'the media' that reprint certain images and embed them in often national narratives of heroism, sacrifice, celebration or outrage. In other words, the images become iconic due to such repeated publication and labelling as 'iconic'. In their seminal study on iconic images, *No Caption Needed*, Hariman and Lucaites (2007a: 27) define photojournalistic icons as:

> photographic images appearing in print, electronic, or digital media that are widely recognized and remembered, are understood to be representations of historically significant events, activate strong emotional identification or response, and are reproduced across a range of media, genres, or topics.

The authors argue that photojournalism makes a vital contribution to public life and by examining a selection of those photographs 'that stand out from all the others over time', the authors move beyond a recognition of their rhetorical power to explore 'five vectors' of influence: 'reproducing ideology, communicating social knowledge, shaping collective memory, modelling citizenship, and providing figural resources for communicative action' (2007a: 6, 9). There is not the space here to unpack how each of these vectors are explored, but Hariman and Lucaites's study has played an important role in raising questions about how certain images come to be recognized as symbolically significant and the role of the media in this co-construction of collective memories about historical events.

There is an emphasis here on emotional identification and the images' role as resources for *citizens* in Hariman and Lucaites's approach. However, their analysis focuses on the images' rhetorical and communicative techniques, generic qualities and circulation across media forms, and does not actually research what the 'public' do with such images: whether they recognize those deemed to be iconic; how they identify with them; and how they talk about them with others. The section below considers how the emergence of digital media practices, along with greater attention to audience spectatorship, has shifted the discussion about iconic images on to how 'publics' actually value, understand and share such images.

From visual icons to viral icons

We have noted how iconic images are commonly defined by their ability to prompt public outrage or discussion following wide circulation across the media. However, as discussed above, this is often assumed by the very same media that repeatedly publish such images and use them as visual shorthand for the most memorable or recognizable aspect of a historic event, often during war or moments of crisis: the raising of the flag at Iwo Jima, the naked girl running from a napalm attack, the fall of the Saddam Hussein statue in Firdous Square. But whereas throughout the twentieth century it was the mass media organizations and political leaders who emphatically promoted certain images, in the age of social media, images can gain momentum across national borders through being shared by citizens rather than authoritative institutions.

Does this mean that different kinds of images now become iconic? How does shareability or 'virality' reconfigure the aesthetics or symbolic qualities of twenty-first century iconic images? Does emotional or moral identification predominate (rather than ideological or national symbolism) in these personalized communication networks?

Alan Kurdi as a global visual icon

One example of a global visual icon is the photograph of Alan Kurdi (or indeed a series of photographs) taken by Nilufer Demir on Bodrum beach in Turkey in September 2015. Alan was one of 12 Syrian refugees who died when their dinghy capsized while attempting to reach Greece. As detailed in a report by the Visual Social Media Lab based at Sheffield University, the images of Alan's lifeless body washed up on the beach and being carried by a Turkish policeman were shared by Twitter users across the world (as far as Indonesia, Malaysia, Lebanon, Spain, United States). Within a few hours, and largely due to certain influential tweeters such as the Emergency Director at Human Rights Watch, Peter Bouckaert, the photographs 'go viral' (Vis and Goriunova, 2015).

The next day (3 September) photographs of Alan appeared on front pages across the world, and by this time non-photographic images also started to emerge (cartoons, illustrations, collages) often as a way to move away from the upsetting realism of the original image. The refashioned images were used to associate Kurdi's death with calls for changes in policies towards the refugees, appeals for charity donations and to make connections with past refugee or humanitarian crises. At times the focus was on Kurdi's *personal* story, at others his image was deployed as a symbol for the *masses* of refugee children or innocents killed due to war. In this way the image became a 'politically oriented' meme (Shifman, 2014), where creative manipulations of the original image are used to comment on the refugee crisis itself or make a link to other issues through layering, replacement of elements or photo-collage (see Chapter 6 for further discussion of memes).

Alan Kurdi's image certainly evoked strong emotional reactions and held the collective attention of citizens from various countries for a few days. The new development for researchers is this ability to track responses through social media and discussion forum posts, where in earlier writing such audience or public reactions have tended to be assumed.

Moral spectatorship on social media

Mette Mortensen and Hans-Jörg Trenz (2016) provide an excellent framework for analysing how social media users respond to and interpret visual icons in their study of reddit discussion groups commenting on the images of Alan Kurdi. Mortensen and Trenz track the emergence of an 'impromptu public of moral spectatorship' in their article and offer a typology of three ideal forms of social media spectatorship: the **emotional** observer who internalizes the suffering of the victim; the **critical** observer who relates their emotional reaction to questions of justice and political responsibility; and the **reflexive** spectator who acts as a meta-observer scrutinizing the media coverage and other users' comments. These are ideal types and the authors acknowledge overlaps, but their argument on the potential for a transnational public sphere constituted by shared spectatorship of visual icons is convincing, even if perceived as momentary in nature.

Although limited to reddit users, the study also turns the spotlight on spectators and their diverse opinions and identifications, rather than de-emphasizing their active role in the meaning-making dynamics of iconic or viral imagery. If public attention, recognition and durability are central to understanding the iconic power of images, we can only welcome more studies that include the voices of the public. Our second case study in this chapter on the image of Omran Daqneesh takes up this challenge and we return to these themes later.

CASE STUDY 1: VISUAL FRAMING OF ISIS PROPAGANDA IN THE UK PRESS

In 2014, an emergent militant group known as ISIS (also known as 'Islamic State of Iraq and the Levant' (ISIL) or Daesh), gained global notoriety when it defeated the Iraqi army in Mosul and began to claim territory across northern Iraq and Syria. Born out of the wars and insurgencies in both countries, ISIS claimed to be establishing a caliphate under its leader, Abu Bakr al-Baghdadi.

ISIS is certainly not the first terrorist group to post videos of executions of their hostages. In January 2002, *Wall Street Journal* reporter Daniel Pearl was kidnapped in Karachi, Pakistan, and a video of his murder by decapitation was released on 21 February, just a month later. One of the demands made by the Al Qaeda affiliated group in the video had been the release of prisoners from Guantanamo Bay, where 'detainees' or 'unlawful combatants' from the War on Terror were being held. Photographs released in January 2002 depicting prisoners in orange jumpsuits soon turned this uniform into a potent visual symbol, adapted by both anti-war activists protesting against the lack of fundamental human rights of the prisoners, and terrorist groups who dressed hostages in similar orange boiler-suits to signify the kidnappings and murders as acts of vengeance against the US-led coalition.

Both US and UK hostages have been pictured in orange jumpsuits in 'demand' videos and in beheading videos. In May 2004, US contractor Nick Berg was murdered in such a manner in Iraq, with the video posted online. Later that year,

British engineer Kenneth Bigley suffered the same fate. This very brief summary is included to show how ISIS were deliberately working with a genre of 'body horror' or gore videos uploaded to sites like Ogrish and later LiveLeak where viewers can watch explicit violence minus the framing of the established news media (Tait, 2008).

In 2014, ISIS posted new beheading videos online, depicting the hostages in orange jumpsuits, but this time in an outdoor desert setting, providing an epic backdrop to the carefully staged executions. Here, we examine the coverage of the executions in the UK press, noting the degree to which the newspapers re-presented details from the video content, and how they variously condemned the brutality of the propaganda images while also picturing the last moments of the hostages' lives in an initially unreflective manner. There was a particular detail which also fascinated the UK media: the masked executioner had a London accent. In their framing of the ISIS beheading videos, the British executioner became the central character – that was, until British hostages were killed.

Research questions and approach

The 'visual framing' approach developed in earlier work by one of us (Parry, 2010a; Parry, 2011), researches the news photograph on three broad levels:

> first, at the compositional level (the content within the frame of the photograph); second, within its immediate news discourse context (the accompanying framing of caption, headline and layout); and, third, across the broader context of a chosen period of time and diversity of newspaper titles, in which certain themes and slants in coverage might be seen to cohere and gain momentum (the broader visual narrative). (2011: 1189)

In this case study, semiotic methods are integrated into a content and framing analysis of press photography in order to examine front-page imagery of the ISIS beheading videos across a range of UK newspapers (*Daily Mirror*, *The Sun*, *Daily Mail*, *Times*, *The Guardian*, *The Daily Telegraph*, *The Independent* and their Sunday equivalents). Visual framing often combines a content analytical approach (recording set features such as size of image, subject gender(s), subject gaze, behaviours or gestures, in addition to stylistic characteristics such as camera angle, distance, focus, picture quality), with a semiotic approach, where such communicative elements are examined for their literal (denotative) meaning, but also their connotative or ideological implications. In other words, by tracing how events and people are depicted and whether certain patterns emerge across the corpus of news images, we can see how a 'perceived reality' is emphasized and whether a certain portrayal of a group of people dominated or was contested. Investigating the images in this way enables the researcher to demonstrate how certain moral and political perspectives are suggested in the coverage, and how this might work to guide readers' attention and understanding of distant events.

In this case, we limit our analysis to front pages of the UK print media and focus on the days when the beheading videos dominated. By comparing multiple news titles, we can note the similarities and differences, in addition to changes over time. With more space available, we would extend this to inside pages over an extended period of time.

In sum, our guiding questions are:

- Do the ISIS videos dominate as front-page news across all print media?
- How does the photographic coverage, along with the captions and headlines, provide a coherent news frame to understand the beheadings?
- How are the main protagonists framed, visually and verbally?
- Do we see changes in approach to the visual coverage over time?

Findings: framing the ISIS execution videos

On 21 August 2014, the front pages of the UK press all led with the news of the ISIS video that depicts the beheading of US photojournalist James Foley, along with the threat to kill another US journalist, Steven Sotloff, two weeks later. However, both the visual presentation and the linguistic framing offer differences. In line with tabloid values, perhaps, it is *The Sun* and the *Daily Mirror* that chose to print full-page images of the masked executioner holding a knife while standing next to the kneeling James Foley in the moments before his murder (see Table 8.1 and Figure 8.1). These are the most dramatic images, capturing the face of Foley just before his death. In this case, the decision to reproduce images of the imminent death of a hostage on their front page raises questions about how newspapers amplify the propagandist message of the terrorists. The *Daily Mail* and *The Guardian* instead show the executioner in close-up, while the *The Daily Telegraph* shows the later moment in the video when Steven Sotloff is presented as the next victim.

The Times and *The Independent* are the only papers not to print the video stills; instead, they print a photograph of James Foley's parents speaking to the media and a map of ISIS-held territory respectively. This variation in visual focus is a notable editorial decision.

With only *The Times* and *The Independent* breaking rank from the dominant visual framing, the uniformity in linguistic framing is notable: all newspapers concentrate on the British jihadist 'Jailer John', with the exception of the aptly named *The Independent*. The nickname may well have come from his captives, but references to the 'Beatles' and 'John' demonstrate how a fascination

TABLE 8.1 Summary of headlines and captions for the newspapers on 21 August 2014.

Newspaper: 21 August 2014	Headline and subheads	Photo caption
Daily Mirror	JOHN THE EXECUTIONER: THE BEHEADING VIDEO FANATIC: He is well-educated, intelligent and possibly from the East End of London: He and his British pals are known as The Beatles by their captives in Syria	CHILLING: Killer holds knife over James in video of sick murder
The Sun	Beheaded by 'The Beatle': SICK NICKNAME OF BRIT EXECUTIONER	Moments before death . . . masked executioner points blade at camera as hostage Foley awaits fate
Daily Mail	FIND THE BRITISH BUTCHER BEHIND THE MASK: Jihadi filmed beheading American journalist is identified as a Londoner known as 'John'	Manhunt: The masked killer from the IS video
The Times	Hunt for 'Jailer John': British jihadist is top target for security services: Fears that journalist's killer will return to UK	John and Diane Foley, the parents of the journalist James Foley. His father said he was haunted by the manner of his murder
The Guardian	Manhunt for a British murderer with hostages' fate in his hands: Experts race to identify 'Londoner': Ex-captive tells of key role in Syria	Security services in the UK and US are trying to identify James Foley's murderer, said to be one of three British jihadists guarding foreign hostages held by ISIS
The Daily Telegraph	ANOTHER LIFE IN BRITISH JIHADIST'S HANDS: Plea to Obama as second US hostage faces beheading: Experts hunt for killer nicknamed 'John of the Beatles'	After beheading James Foley, a US journalist, a jihadist who spoke with a British accent is filmed holding Steve Joel Sotloff. A translation of the Arabic subtitle above reads: 'The life of this American citizen, Obama, depends on your next decision'
The Independent	Where do we go from here?	An American victim. A British killer. A region in turmoil. A Western world united in outrage but bereft of ideas. As the Prime Minister cuts short his holiday for crisis talks and President Obama joins the chorus of condemnation, the murder of US journalist James Foley has made one thing brutally clear: the relentless rise of the self-styled 'caliphate' in Iraq and Syria is a problem the UK can no longer ignore

Source: A selection of front pages can be viewed at: www.thepaperboy.com/uk/2014/08/21/front-pages-archive.cfm

with the Britishness of the killer overshadows the loss of life as a 'manhunt' narrative unfolds. While the moral judgement and apportioning of blame are clearly attributed here, the popular culture references, both visual and verbal, focus on 'Jihadi John', and construct him as a Hollywood-style 'baddie'. This frames the murders through a dramatized and nationally focused lens where, in its most tabloid form, the human consequences of this act of political violence are diminished in favour of co-constructing a pantomime villain who revels in his own brutality (Figure 8.1).

The horrific regularity of the videos released by ISIS continued over the next few months. On 3 September 2014, frame-grabs from the beheading video of Steven Sotloff again appeared across the newspapers, with *The Sun* printing a pixellated image of the next

FIGURE 8.1   *Daily Mirror* front page on 21 August 2014 (all front pages can be viewed at: www.thepaperboy.com/uk/2014/08/21/front-pages-archive.cfm).

hostage threatened. This time it is to be a British aid worker, David Haines. The newspapers had already visually replicated the demands of 'John' by concentrating on his direct gaze and address to the camera through his black balaclava-style mask. Now they also quote his words directly as headlines, with the *Daily Mirror* even presenting them without the use of distancing quote marks: 'I'm back . . . and I'll kill a Brit'. Depicted holding the knife up and talking to the camera, the British jihadist appears to set the disturbing and uniform media agenda here. Within our sample, *The Independent* alone chose to print a portrait of Sotloff from before his capture (see www.thepaperboy.com/uk/2014/09/03/front-pages-archive.cfm for a selection of front pages on this date).

However, when David Haines was murdered in the same manner, we notice fewer newspapers choosing to show images from the video on their front page: only *The Sun on Sunday*, *The Sunday Telegraph* and *The Sunday Times* reproduced the now eerily familiar images on 14 September. It is worth noting that the sister Sunday versions of these newspapers operate under different editors from the weekday equivalents. By 25 September, in response to calls to do something to halt ISIS, Prime Minister David Cameron announced that Parliament would be recalled from its summer recess to vote on conducting air strikes in Iraq. Clearly, the distinctive visual address of the videos and the news framing worked to boost a sense of moral certainty and to legitimate military intervention. The military action gained support across the political parties and Parliament voted overwhelmingly in favour of joining US-led airstrikes on 26 September 2014.

On 4 October, reports of the killing of the second British hostage, taxi driver and aid worker Alan Henning, revealed a different approach by many of the newspapers (see www.thepaperboy.com/uk/2014/10/04/front-pages-archive.cfm). The beheading was front-page news across the sample, except for the

FIGURE 8.2 *The Independent on Sunday* front page on 5 October 2014 (all front pages can be viewed at: www.thepaperboy.com/uk/2014/10/05/front-pages-archive.cfm).

Daily Mirror which led with a celebrity-focused story. Within our sample, only *The Daily Telegraph* used the ISIS video footage as its main image; all the others printed photos of Henning either in the aid convoy or smiling to the camera as he holds a refugee child during his volunteer work, with the *Daily Mail* printing the two images alongside each other. Five out of the seven newspapers used the term 'beheaded' in their headlines, with only two out of seven naming a perpetrator: ISIS. This was in stark contrast to the earlier personalized focus on the executioner 'John'.

The following day, *The Independent on Sunday* printed a now-famous front page, almost entirely in black, with the words: 'On Friday a decent, caring human being was murdered in cold blood. Our thoughts are with his family. He was killed, on camera, for the sole purpose of propaganda. Here is the news, not the propaganda' (*The Independent on Sunday*, 5 October 2014). This blacked-out front page makes a powerful statement about the complicity of news outlets in reproducing the terrorist videos (Figure 8.2). With terror attacks conducted by those affiliated with ISIS in cities such as Brussels, Paris, Nice, Istanbul, London and Manchester, the media debate shifted pointedly to denying the terrorists the media space they so desired. In France, a number of media organizations have since decided to no longer print the photographs or names of terrorist suicide attackers to deny them 'posthumous glorification' (Borger, 2016).

Case study conclusion: the allure of the masked executioner and why it matters

In summary, we can note a striking uniformity in the front-page prominence of the video-led story, and in the visual and verbal framing of ISIS's actions. The visual coverage served to highlight the otherness and brutality of the executioner, while at the same time the newspapers were drawn towards his mysterious identification as a 'Brit'. The casting of the hooded figure as a fanatical terrorist is not surprising, but the choice to emphasize and personalize 'Jihadi John', in both visual and linguistic modes, arguably diminishes the human dignity of the murdered James Foley. In prioritizing and dramatizing 'John's' demands (visually conveyed in his audience address via the video-stills), the print media allow ISIS to set the agenda and the terms of the public debate.

Over a period of less than two months, the newspapers significantly adjust the predominant visual framing of the ISIS beheading videos. There are a number of possible explanations for this, including: the significance of the nationality of the victim; the shifting newsworthiness and declining shock factor as the video releases take on a disturbing regularity; and the apparently dramatic shift in the government's military policy towards ISIS, in some cases cheered on by the media. But this visual shift over time also indicates a growing sense of self-reflexivity among the news media, with a refocusing of attention on to the victim's life, and a clear de-emphasis on 'Jihadi John' and the captured moment of violent death. This is not the place to determine exactly how each of the factors combined to shift the national news agenda, especially as we cannot take into account other relevant developments in the conflict here, but the case study does offer insights into the alluring power of such images of 'body horror' and how an initial mode of fascination and dramatization gives way to a more humanized and reflective style of reporting.

This is an image-led news story, initiated by a militant group determined to visualize their ruthlessness. The choice of front-page image is undoubtedly a crucial element in the editorial decision-making process and contributes to the interpretive news frame on offer. The nature of such coverage is illustrative of the ongoing journalistic struggles between the duty to bear witness and the obligation to resist the terrorizing narrative of violence and vengeance.

ACTIVITY

Table 8.1 above presents a summary of the headlines and captions for the front pages on a single day. We only have space to include a single day in detail here and we are unable to reprint all the images, but there is much potential here to further analyse the various communicative modes in the news framing in order to demonstrate how a nationally focused perspective leads to an emphasis on 'Jailer John' in this coverage. What do the selected words, images and the emphasis on certain details tell us about the favoured framing of the ISIS beheading videos? How is the threat to UK citizens or others imagined?

Take your own selection of front pages on a day of a dramatic news event and conduct a similar analysis – for example, compare how newspapers from different countries visually report on the same event and how such images are sourced, if credited. If there is standardization across coverage, where do the differences break through? Is there enough difference to offer a counter-frame to the dominant form of coverage?

A useful full-length study to consult is Shahira Fahmy's (2007) article on the visual framing of the toppling of the Saddam Hussein statue in Baghdad in April 2003, as reported across 30 countries. You can use the Paperboy archive to find the front pages using the date within the following URL template: www.thepaperboy.com/uk/YYYY/MM/DD/front-pages-archive.cfm (also swapping the country 'UK' for other countries such as 'US or 'Australia').

CASE STUDY 2: OMRAN DAQNEESH: THE BOY IN THE AMBULANCE

Our second case study reports on a short survey conducted with students enrolled on the 'Visual Communication' class (2016–17) at the University of Leeds. This study considers how citizens respond to certain images that become 'visual icons' – in this case, the image of Omran Daqneesh, a 5-year-old boy pictured in an ambulance in Aleppo after he was pulled from the rubble following a bomb strike on 17 August 2016. Omran's image appeared almost exactly a year after the photos of Alan Kurdi's death had gone 'viral'.

Omran's photograph was taken by Mahmoud Raslan and was accompanied by video footage by Mustafa al-Sarout, uploaded to the Aleppo Media Center's YouTube channel. In contrast to Alan Kurdi, Omran Daqneesh was rescued and placed in an ambulance, and so could be viewed as a symbol of survival. However, as we shall see, it is more complicated than that.

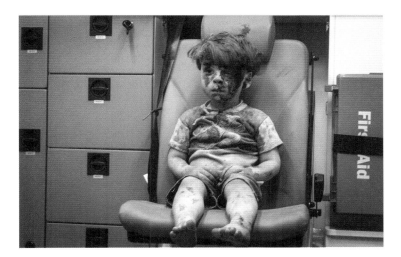

FIGURE 8.3 Image of Omran Daqneesh widely circulated in August 2016. Copyright: Mahmoud Raslan, Andalou Agency/Getty Images. Reproduced with permission.

Research questions and approach

The questionnaire used for this case study first collects demographic data and asks respondents about their interest in politics, news media preferences and their emotional connections to the Syrian war more generally, before turning to the image itself. There were two versions of the questionnaire: one displayed the photograph (Figure 8.3) without a caption, and the second included a caption, copied from the headline for an online NPR article written by Malaka Gharib, which asked: 'The Little Boy In Aleppo: Can One Photo End A War?' (2016). Other than this variant, the questionnaires were identical.

Eighty-seven students were randomly assigned version 1A or version 1B of the questionnaire: 42 completed '1A' with the caption and 45 completed '1B' without the caption. We were interested here in whether the caption, 'The little boy in Aleppo: Can one photo end a war?', helped to aid memory or stimulate a strongly expressed response. While the caption questioned the image's 'power' to end the war, it also provided factual information that could have worked to 'ground' the photograph. Crucially, the questionnaire was designed to gather **affective responses** in addition to testing contextual knowledge, political and cultural associations.

Such a study conducted with a narrow demographic of students cannot generalize any findings about the way in which news photographs are consumed by the general public, but it can point to the formative role played by prior attitudes and feelings towards the war and the *degree* to which photographs are perceived to communicate a certain message or invoke an emotional response.

So, here we summarize some of the guiding research questions that have informed the design of the questionnaires:

- How recognizable is the image?
- How uniform or varied are the students' responses to the same photograph?
- How do the respondents *express* their political and emotional responses to the image?

There is not enough space to report on the full range of findings here, so this is a selective discussion, designed to engage with the existing research on visual icons and offer potential lines of enquiry for similar studies. The full questionnaire can be requested from the authors.

Findings: initial survey results

It is perhaps worth reiterating the limited nature of this study when it comes to the particular cohort of student respondents, 95 per cent of whom were in the 18–24 age category. Of those who completed the study, 72 per cent were British or with dual nationality including British; the others came from a wide range of countries, mostly European but with most of the global regions represented by one or two students. The vast majority were female (78 per cent) and so with only 18 male respondents (plus one with no gender given), comparisons between female and male participants are also limited.

Our main task here is to reveal how audience surveys can be employed to interrogate some of the assumptions made about the 'impact' of iconic images and especially those depicting children living in violent circumstances. When asked if the Omran Daqneesh image was already familiar, across both survey versions 83 per cent of respondents said 'yes'. This was slightly higher for those who also viewed the *uncaptioned* image (1B: 89 per cent), as opposed to the image with the caption (1A: 76 per cent). This is perhaps surprising, as we might expect the text to aid recall and help to place the image in context. However, with such small numbers this is possibly just an incidental outcome of the distribution of the questionnaires.

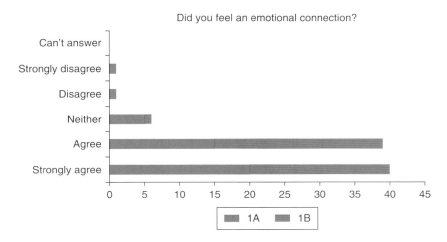

FIGURE 8.4 Responses to the statement, 'I feel an emotional connection and reaction to the photograph'. Numbers above refer to responses in each category, not percentages.

Asked if they felt 'an emotional connection and reaction to the photograph', overall 91 per cent either 'agreed' or 'strongly agreed' (Figure 8.4). Pre-existing political interest and familiarity with the image appear to correlate with the tendency to express a stronger emotional response to the image of Omran, whereas the presence of the caption did not. What is striking here is the almost universal agreement on its emotional power and the high numbers who 'strongly agree'.

The survey shows that large numbers of respondents were both familiar with the image and recognized an emotional connection. When we asked respondents further details on the nature of the emotional response, the most commonly 'agreed' to were sadness, disgust, anger and frustration (Figure 8.5).

Respondents were most likely to disagree with feeling proud, hopeful, secure and indifferent. Whereas emotional responses were relatively consistent, there was less agreement on what the image might mean for the rights and wrongs of Western military intervention. So, while 93 per cent agreed that the image

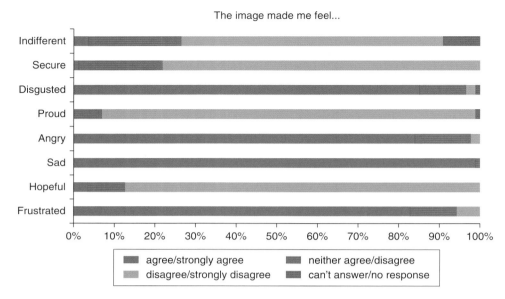

FIGURE 8.5 All questionnaire (1A and 1B) responses to the statement, 'The image made me feel . . . '.

'communicated something powerful about the war', we see a much more mixed response to the statement, 'I feel that the photograph supported Western governments' (e.g. US/ UK/ France) justifications for military intervention' (Figure 8.6).

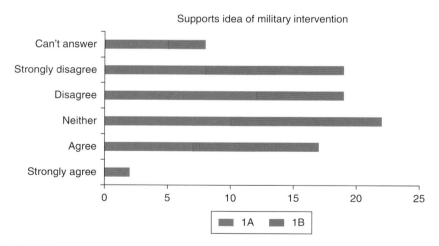

FIGURE 8.6 Responses to the statement, 'I feel that the photograph supported Western governments' (e.g. US/ UK/ France) justifications for military intervention'. Numbers above refer to responses in each category, not percentages.

While 44 per cent expressed disagreement or strong disagreement with the idea that the image might support justification for military action, 22 per cent agreed with the statement, with a high proportion selecting 'neither' or not answering (34 per cent). The respondents broadly agreed on the image's communicative and emotive power, but not on its role in the more complex foreign policy implications. As the later 'open text' answers reveal, this is partly due to a self-perceived lack of knowledge about the war, and points to an uncertainty and complexity that can sometimes be de-emphasized in media commentary, where the image is cast as an instantly decipherable visual icon with the power to 'end the war'. The above responses are restricted to the pre-set questions, so we also included open text questions to explore the prompted memories and responses in more detail.

Open-text comments from respondents

In addition to the closed survey questions discussed above, the questionnaire included the following open questions or space for comment:

- Please describe in your own words what you know about the circumstances for the image (e.g. the name of the child, the place, when it appeared, etc.), even if this is only a vague memory.
- Does it bring to mind other images of children from war? If you can only describe them roughly, that is fine.
- Please write any other reactions or impressions you had relating to the photograph that you would like to share.

The intention here was to gather the students' responses in their own words and this gives us a richer sense of how respondents encounter the image (and others like it), the nature of the information they tend to recall and the cultural resonances such images hold. Only two respondents left no further comments at all, and a common reaction was further expression of sadness, anger and feelings of impotence or hopelessness when it came to actually having the power to effect meaningful change.

While some respondents reflected on their own limited knowledge, in fact the answers reveal a deeply engaged and empathetic set of responses.

The first question, which asked students to describe what they remembered of the image in their own words, prompted answers most closely related to the '**denotation**' of the image (its literal meaning), but also the **connotations** where respondents referred to details that often required connections to cultural meanings beyond the image and caption (i.e. how the values or interpretations of this image might connect to others). The second question further encouraged reflections beyond the immediate image, to gauge how their recollections on this image provoked memories of other war images of children, and therefore offered insight into the wider cultural resonances of iconic war imagery for this group of students. In Barthes' ([1961–73] 1977) semiotic model, this would relate to the function of '**myth**', where decoding such images helps us to make sense of the particular worldview on offer and its ideological significance. The third question was very 'open', so it was interesting to see how respondents made additional contributions and further questioned the mythic power of the image through various articulations, including personalized, emotional reflections and attributions of responsibility (in terms of both political culpability and media ethics).

It is possible, then, to provide a collective semiotic reading of the image (or the image plus caption) from our respondents, summarizing the most frequent answers and recollections. The key details offered by respondents were that the picture depicted a little boy in the back of an ambulance, rescued from an airstrike in Aleppo, Syria, and that his home had been destroyed in an airstrike. The additional details proffered are of interest here: some respondents markedly pointed out that they did not know his name; others remembered *when* they saw the image but gave no other details ('I don't know the name of the child, where the child is supposed to be and it happened in August 2016'); his aloneness is remarked upon and also that other members of his family died (his brother). In fact, only one respondent offered a first name – 'Omran'.

The idea that photographic images hold more power than moving images (Sontag, 2003) is perhaps challenged by the remarks that indicate that they saw it on Facebook as a video, rather than a still image. Indeed, some point out the boy's *movements* as an aspect that makes the image so shocking: 'He seems unphased (*sic*) by what had just happened, and actually touched his bloody head, still not reacting to his injuries – I saw this video on Facebook'. Although many respondents mentioned airstrikes or bombing, few offered information about who might have dropped the bombs, with one answer demonstrating the difficulties in attributing responsibility for such actions in the Syrian war: 'carried out by either British, American, Russian or the Assad regime'.

With regard to other images of children from war that the image recalls, perhaps not surprisingly the images of Alan Kurdi's 'lifeless body' washed up on the beach are the most frequently mentioned. The next most stated is Nick Ut's napalm attack image from Vietnam, although this could be due to the fact that it had been widely circulated and discussed on Facebook earlier that same month. This circulation followed a controversial decision by Facebook to censor the image because it featured a naked child. Facebook backed down after initially defending its 'Community Standards' policy, but not before some commentators expressed concern over its 'abuse of power' as a major news distributor (Levin et al., 2016).

With limited space to explore such a rich data set, we highlight three observed patterns:

- Personalized reflections and empathetic imagination:
  - 'Makes me feel sad as I think how I'd feel if that were my younger brother/ family and that situation.'
  - '. . . feeling impotent is the major reaction/feeling I have when seeing such images. *why*. Why is it like this?'
  - '. . . I wish we as mature university students with open minds could learn more about the agendas of governments.'
- Attribution of blame or responsibility:
  - 'Western governments are not being careful enough about avoiding civilian casualties. There can be no "winning side" – both "sides" are causing irreparable damage and casualties/trauma.'
  - 'Pure shock that nothing is being done to help these innocent people and all the government is concerned with is Brexit and saving its own skin.'
- The role of the media and the nature of coverage:
  - 'I feel frustrated for the amount of images displayed on TV in this way. We are programmed to feel sorry for the subject in the image but for me frustration occurs.'
  - 'The selectivity of the news shown in the UK suggests to me that Syria is often forgotten about. While Paris attacks and Kim Kardashian being robbed at gunpoint becomes **front-page** news. Thousands of people are dying in Syria but yet it is not "important enough" for the front cover.'
  - 'It's extremely exploitative.'

The three categories above were those that we inductively determined as prominent patterns of response after reading through all the open text comments. However, there is an interesting overlap here with Mortensen and Trenz's (2016) findings from their study of the reddit comments on the Alan Kurdi photograph discussed earlier. The patterns of response closely map on to the three modes of spectatorship outlined earlier on in this chapter: the **emotional** observer who internalizes the suffering of the victim; the **critical** observer who relates their emotional reaction to questions of justice and political responsibility; and the **reflexive** spectator who acts as a meta-observer scrutinizing the media coverage and other users' comments.

Case study conclusion: the potentials and limitations of moral spectatorship

Although conducted in a very different context of a survey instead of online discussion, our results support Mortensen and Trenz's typology of moral spectatorship. Whereas a small number of respondents claimed that their generation had become desensitized to such images, this perspective was countered in much larger numbers by those stressing their sadness, shock, frustration, guilt and helplessness. In some cases, this was expressed through imagining the boy's plight as their own or a family member's, but in other cases Omran stood for all the other children affected by war. Whether *personalized* or *generalized* in such articulations, the emotional responses indicated a desire to question the circumstances of the war, but at the same time an overwhelming impotence in knowing how to help: 'Horror, deep sadness and an immense feeling of inability.'

Given that so few had taken part in activist activities (only 26 per cent had signed a petition related to the Syrian war and only 3 per cent had attended a march or demonstration), this does raise questions about feelings of political efficacy among this group, and possibly connects to wider concerns about the political engagement of young people. The political ambivalence and perceived lack of knowledge demonstrate that such images from conflict encourage critical and self-reflective questions, but cannot be expected to equip spectators with all they require to understand the complexities of the war.

Conducting analysis of such surveys or online discussion groups allows researchers to move beyond assumptions of how spectators react to and remember 'iconic' images, and shows how audiences make sense of images in different media contexts. Where the literal or denotative meaning remains relatively stable in meaning across responses, and its emotive power was acknowledged by many, what the image signifies in terms of the rights and wrongs of military intervention and how governments might 'end the war' is harder to determine.

CHAPTER CONCLUSION: THE BURDEN OF THE WAR IMAGE

We should be vigilant, then, for any claims that certain pictures 'change the world' or 'end the war' as this over-simplifies the complex geopolitical and historical contexts that create the circumstances for particular images. It also means paying attention to the authoritative framings and media platforms within which they circulate and through which their cultural significance and symbolism are shaped. However, digital networks of image circulation and amateur imaging practices have possibly disrupted the traditional co-creation of 'iconic' images by the political and media elites, as citizens also contribute to image creation, public distribution and the formation of shared cultural memories. Social media technologies also facilitate research into 'audience' reactions and remind us that spectators' affective and moral responses are varied and complex.

This chapter has focused on the print media's coverage of acts of terror and audience survey responses to a 'digital visual icon'. It has engaged with two of the key debates about the role of images in warfare: how media potentially become a propaganda tool for terrorism or military intervention; and how the 'iconic' innocence of suffering children prompts emotionally complex and politically varied responses. But these are just two brief enquiries into the war image, and other significant questions require our continued critical attention:

- How do mainstream media incorporate images from 'citizen witnesses' and is this having a broader impact on the way war is visualized?
- How do war photographers cope with the dilemmas of picturing people at their most vulnerable?

Each new war prompts scholarly interest in how the patterns of visualizing conflict draw upon established symbolism and mythic pasts, at the same time as finding new shifts in the aesthetic characterization of war. The dangerous work of war photographers and citizen witnesses is often the first stage in the process of revealing injustices and suffering, as well as heroism and sacrifice. It therefore remains essential to investigate both the ethical dilemmas of picturing people in the most dangerous conflict situations and how the resulting images are believed to evoke moral or political responsibilities.

FURTHER READING

The scholarship on media and war is vast, with growing attention to the relationship between war and imagery resulting in a rich collection of relevant works. Short essays on topics such as icons, trauma and violence are available in *Visual Global Politics* (Bleiker, 2018). The SAGE journal *Media, War & Conflict* is also a good place to search for articles on visual coverage of war.

PART III
COMMODITIES

9 The visual attractions of advertising and promotional culture   185

10 Visualizing lifestyles as commodities   209

11 Brands as visual experiences   233

9
THE VISUAL ATTRACTIONS OF ADVERTISING AND PROMOTIONAL CULTURE

The next three chapters come under our general theme of 'commodities'. The word 'commodity' implies something that can be bought or sold; it has economic value and so is associated with the realm of commerce. But commodities also have symbolic and social meanings. In promotional practices or the marketing of goods, we can observe how a complex relationship between our identities, values and the things we buy is cultivated and sustained. In a sense, then, our discussion of 'commodities' builds upon our preceding strands of 'identities' and 'politics'. The materials used to promote products or services also provide rich cultural artefacts through which we can discern some of the dominant aesthetic, moral and political ideals in the times of their production.

This chapter considers the history of images and commodity culture as a key concern for media scholars. As we continue to consume media in novel formats and via new devices, advertisers and promoters find new ways to attract our attention and convince us of the pleasures that await us.

This chapter:

- introduces how promotional culture, advertising and capitalism sustain each other;
- explores the history of critical analysis of advertising and promotion;
- explains how visual and semiotic analysis help to reveal the codes and conventions of advertising and how certain ideologies become naturalized in such codes;
- conducts an analysis of the O2 'Be more dog' advertising campaign to explore the appeal and humour of animals (Case study 1);
- reveals how film trailers combine rhetorical appeals of genre, story and stardom in their announcements for 'coming attractions', illustrated with an analysis of the *Black Panther* (2018) trailer (Case study 2).

ADVERTISING AND PROMOTIONAL CULTURE

Who isn't involved in promotional culture these days? As academics, we are encouraged to promote our publications and talks through social media and university communications advisers. In our personal lives, many of us promote a version of ourselves on Instagram and Facebook (see Chapter 3). For public sector and commercial institutions, reputations are managed by public relations consultants keen to secure favourable publicity, while politicians hope to perform the right kind of visibility to persuade voters of their leadership qualities (see Chapter 6).

Aeron Davis argues that we live in promotion-saturated times. Working with Andrew Wernick's coining of 'promotional culture' in 1991, Davis applies a sociological approach to the evolution of the promotional industries in his book

Promotional Cultures (Davis, 2013; Wernick, 1991). The use of the plural 'cultures' by Davis here suggests how such promotional imperatives and techniques have been adopted beyond the advertising, public relations, marketing, lobbying and branding professions, to reshape the work of other organizations – whether charities, social movements, political parties, or indeed any sector of society.

When we pay particular attention to the links between advertising and the cultural industries, the cross-cutting nature of promotional materials becomes awe-inspiring, and this is where we see the different industries reinforcing each other – music, films, merchandise, dolls – each endorsing and sustaining the other. As Wernick (1991: 121) eloquently writes:

> It is as if we are in a hall of mirrors. Each promotional message refers us to a commodity which is itself the site of another promotion. And so on, in an endless dance whose only point is to circulate the circulation of something else.

In this 'endless dance', Wernick suggests self-perpetuating and dizzying exchange, not only of commodities, but the signs, values and symbols associated with them. Indeed, advertising and media industries have a particularly symbiotic relationship – newspapers and magazines rely on advertising revenue, while from the mid-twentieth century onwards the rise of commercial television was dependent on the boom in advertising (Davis, 2013). Despite the more recent ability to fast-forward adverts on television, to 'skip ads' at the start of YouTube videos and use ad blockers on website browsers, advertising strategies adapt with media uses and the industry retains some degree of its economic and cultural hold.

Nicholas Holm's (2017) critical introduction to advertising and consumer society offers a superb resource for those wanting to find out more about advertising's role as both an essential part of our capitalist economy and as an artistic form. In his introduction, Holm sets out five reasons to study advertising: advertising is everywhere; advertising is weird; advertising is where the money is; advertising is beautiful, inspiring and entertaining; advertising is political. This short list hints at how Holm interrogates the pervasive and little understood supremacy of advertising as an industry, but also recognizes its creative and political potential.

As Holm writes, advertising is so ubiquitous that we sometimes forget just how bizarre it is – whether in its surreal imagery or as a wider practice. Judging its 'effectiveness' is almost impossible and yet it is a huge global industry, thought to keep other industries afloat and so provide employment for many. For media and cultural studies scholars, adverts have often been approached with a suspicious attitude, undergirded with a concern for the reinforcement of a capitalist ideology based on false or unattainable promises, and yet advertising has a wider cultural and artistic role. As Holm argues, approaching advertising as an **art form** leads us to think about those making adverts 'as passionate, creative people with different politics and who want to tell stories and create meaningful impressions and interventions in the world' (2017: 8). But he also explores how 'creativity' can become a problematic term, where the appeal of working 'creatively' leads to overworked and undervalued staff whose artistic inspiration is exploited to serve the demands of the **dominant capitalist ideology** (2017: 168). The above summary points to the sometimes contradictory appeals of advertising, and suggests a rich and challenging wealth of material for visual communication researchers.

The image in promotional culture or why semioticians love advertising

As detailed in Chapter 2, Roland Barthes' semiotic analysis of a Panzani pasta advert remains one of the most influential analyses of advertising. In his essay, 'The Rhetoric of the Image', Barthes ([1961–73] 1977) outlines the 'three messages' of the advert: the linguistic message, which usually works to 'anchor' the meaning; the coded iconic message (visual connotations); and the non-coded iconic message (the 'literal' objects depicted or the 'denotative' sign). In examining the advert, the analyst moves between these layers of communicative expression – linguistic, non-coded and coded signs – to recognize how meaning is created in the interplay of such messages. Adverts in particular are designed to associate certain values with their products (luxury, glamour, youthfulness, joy), so this process of meaning-construction is often particularly visible in this cultural and promotional material.

There is an abundance of sources for both the original essay and later readings of Barthes' analysis, so we will not rehash this important contribution in detail here. Suffice to say that Barthes' application of semiotics (or the science of signs) to a range of popular cultural artefacts demonstrated how the visual or 'iconic' message was also worthy of systematic and close examination. As we discussed in Chapter 2, this early structuralist semiotic approach is not without its difficulties and limitations, especially when it comes to acknowledging cultural differences, but it paved the way for a serious appreciation of how visuals perform certain communicative or rhetorical functions. For advertising in particular, which has an explicit intention and *persuasive* purpose, a semiotic approach underscores the 'work' of **codes** and **conventions** in understanding the advert's communicative appeal.

There is a broader concern underpinning this kind of analysis: the way in which certain ideologies become naturalized in codes of advertising. One overarching consideration here is how advertising is inextricably linked to capitalism. The industry is all about encouraging people to buy products made by other industries. The viewer is addressed as a consumer, albeit by many different and clever strategies – some direct, others opaque. However, the relationship is also more complex than that, as Marita Sturken and Lisa Cartwright (2017) explain in their discussion of how we construct our identities through the consumer products central to our lives. By this understanding, we do not simply buy products, but convey our personalities through the clothes we wear, the cars we drive, etc., and so advertising encourages us to think we assume the qualities of the desirable people depicted in such adverts just by buying their products.

Much of the critique of advertising and wider consumer culture is influenced by the Frankfurt School theorists, such as Theodor Adorno and Max Horkheimer ([1944] 1993), whose essay on 'The Culture Industry' captures their fears of a mass entertainment industry (Hollywood, music, magazines), ruled by the requirements of the capitalist production line where the smooth supply of reliable goods leads to homogeneity, conformity and banality. Adorno and Horkheimer also point out the paradox at the heart of an industry that promises fulfilment but can never quite satisfy, and so advertising and the culture industry merge as one fuels the other (the film star on the billboard, the product placement in a film), and the cycle repeats. They conclude the essay: 'The triumph of advertising in the culture industry is that consumers feel compelled to buy and use its products even though they see through them' ([1944] 1993: 24). This famous sentence is significant in that it hints at a less passive role for the consumer than has often been assumed in the Marxist writings of Adorno and Horkheimer.

The point here is that consumers are aware of being manipulated, but they continue to buy and use new products, making choices between brands based on dubious appeals. This could apply to a range of products, not just those deemed part of the culture industry. Consumers 'see through' marketing ploys, but might still find an attraction or affinity, when choosing one product over another despite their similarities. We are all complicit in this consumerist ideology despite our knowing that its promises are unattainable. This is not simply about pushing us to buy certain products; it is about the reinforcement of aspirations for a better world and better life 'shaped by the promises of advertising' and the ideology of consumerism (Holm, 2017: 90).

Visual content analysis studies of advertising

We can see how the concerns about advertising emerged during the twentieth century in particular and how its powerful role in society was theorized and challenged. Much of the above writing tends to assume that advertisers possess the skills for powerfully persuasive messages. If we reserve judgement for now about whose interests are served in advertising, we can refocus on to the strategies and techniques employed. For industry experts, such techniques of persuasive communication are valued for their attractiveness and effectiveness, but critical scholars examine the same techniques for the ways in which they enact and perpetuate certain social roles (such as in stereotypical depictions), and certain values and desires (a narrow understanding of beauty aligned with happiness and fulfilment).

Classic studies on advertising include sociologist Erving Goffman's *Gender Advertisements* (1976), in which he examined the way female and male bodies were positioned in adverts in what he calls the 'ritualization of subordination', where men's higher social status is symbolized in their higher positioning in adverts. In contrast, women's smiles, gestures and stances, such as the 'bashful knee bend', can be 'read as an acceptance of subordination, an expression of ingratiation, submissiveness, and appeasement' (1976: 46).

While Goffman's study was well illustrated and highly influential, it could be argued that he selects the adverts that support his argument and categories. A modern-day content analysis study would require a much more systematic approach to its **sampling method** in order to provide evidence that such depictions predominated and to reliably track changes over time (see Chapter 2). Indeed, Goffman's study has inspired a number of content analyses, both testing his categories and pursuing further concerns of sexualization, objectification and sexual agency (Bell and Milic, 2002; Gill, 2008). Indeed, there is a continuing fascination with the ways in which feminine and masculine identities are constructed in adverts and other promotional material such as film posters or magazine covers (which arguably 'advertise' their contents in a purposeful manner). As Schroeder and Zwick (2004: 22) demonstrate, the forms of masculinity and femininity promoted in advertising campaigns work with the logic of consumerism where 'normative sexual dualism', or an emphasis on a 'system of difference', is reinforced and maintained:

> Within this system, iconic masculine activities such as shaving the face, driving fast cars, having a hearty appetite, smoking cigars, and drinking liquor are juxtaposed to feminine visions of applying makeup, driving a minivan, eating "light," doing the laundry, and decorating houses.

The above is almost parodic in its summation, but instantly recognizable. Idealized images of male and female bodies continue to be predictably dominant in promotional messages, but are also mimicked or subverted in more playful advertising strategies.

The available research on gender representation in advertising is vast. Such research cuts across the practice-focused marketing, consumer and advertising journals, to more critical traditions of discourse and gender studies journals. Due to the abundance of such studies, and the fact that we examine gender representation in a number of the other chapters, here we take a swerve away from this dominant focus, to concentrate first on animals in advertising in our case study below.

Why are there so many animals in adverts?

Once you start looking, you'll notice just how often animals appear in adverts – and not just those for pet food, but as symbols, brands and fantasy figures (Brown and Ponsonby-McCabe, 2014). Intriguingly, it is sometimes possible to see gender, sexuality, social class, age, ethnicity markers and other stereotypical logics underlying the way in which animals are depicted in adverts. It could even be claimed that the added layer of fantasy allows for questionable stereotypical formulations in the animal characters, which might otherwise attract negative attention if in human form.

Perhaps one of the most successful characters in a long-running campaign is Aleksandr Orlov, the animated meerkat protagonist in the comparethemarket.com adverts. A price comparison website, the central joke of the campaign is a misunderstanding between the website names 'comparethemeerkat.com' and 'comparethemarket.com'. This confusion is enhanced by the faux Russian accent and a backstory that plays on Orlov's oligarchic lifestyle in his mansion. The CGI animated version of the meerkat characters traded on the global popularity of the southern African animals following a number of innovative natural history documentaries in the 2000s which showcased their cooperative behaviour in the wild (as well as their less appealing habits).

Giving an African animal a Russian accent makes no sense, unless you recall the weirdness of advertising as mentioned by Nicholas Holm (2017) and cited earlier. The popular appeal of a wild creature is here adopted by an advertising agency, CGI animated, given an East European accent and backstory, used to sell a website that has no material product and then turned into a commodity as a cuddly toy for sale via the website – and that's not to mention 'Orlov's' best-selling autobiography released in 2010.

So what makes animals so appealing to advertisers? What properties of the animals are being conferred on to the product in these adverts, if it's even that simple? In his essay, 'Why look at animals?', written in 1977, John Berger writes that what distinguishes humans from animals is our capacity for 'symbolic thought' and yet almost universally 'the first symbols were animals' (Berger, 1980: 9). In origin stories and myths from around the world, animals feature 'as messengers and promises' (1980: 4). Despite farming animals for meat, they are given magical functions in stories: 'They were subjected *and* worshipped, bred *and* sacrificed' (1980: 7).

The strong distinction between 'human' and 'animal' (and the implicit superiority of the former) is an assumption questioned by some, and a growing interdisciplinary field of 'animal studies' seeks to better understand human–animal relations and how humanity is defined in relation to animals (Waldau, 2013). The fascinating work of this developing research strand is beyond the scope of this chapter, but we hope that thinking seriously about animals also comes with a capacity for greater compassion towards them. How do images of animals prompt us to feel differently about them? By projecting human qualities on to animals (**anthropomorphism**), are we actually more interested in our imagined similarities than the animals' own merits? And how does this relate to the weird world of advertising?

CASE STUDY 1: ANIMALS IN ADVERTISING: O2'S 'BE MORE DOG'

O2's 'Be more dog' campaign ran from 2013 to 2016, winning many plaudits from the industry, including 'campaign of the year' in 2013. The multi-media campaign was developed by VCCP for Telefonica (who own O2), apparently helping the mobile communications company to move from fourth to first in market share, according to the agency's website (www.vccp.com/about-us/).

The original adverts were voiced by actor and comedian Julian Barrett and used rock band Queen's *Flash Gordon* as its soundtrack. The campaign skilfully combined the viral power of the cat-obsessed internet with an advertising strategy that quickly attracted press attention: as seen in the *Daily Mirror* report that the launch advert had been watched 385,000 times on YouTube in its first 48 hours (Evan, 2013: this online article also contains the full YouTube and TV advert). So, in addition to being available across billboards, YouTube, social media, newspapers and television adverts, the campaign attracted the kind of word-of-mouth attention that money can't buy. 'Be more dog' benefited from having a positive life-affirming message: in this case, the message of being enthusiastic in life was intended to translate into using more of O2's services. The social media campaign also won awards, using innovative dual screen technology to 'throw' a virtual Frisbee from a phone to a laptop, encouraging users to share 'dog bombs' across Twitter and Facebook (personalized video messages that encouraged their cat-like friends to 'be more dog') and even handing out Frisbees in parks to create photo opportunities to share on social media (Shorty, 2013).

Research questions and approach

As others have observed in relation to a semiotic approach, here we want to interrogate the 'how' of persuasive communication, or its 'poetics', more so than its 'politics' (Hall, 1997: 6).

In any analysis of this kind, we can ask three questions:

- What is the advert selling or promoting in the most basic terms?
- What is the advert selling in symbolic or abstract terms?
- How are qualities of the symbolic meaning coded into the advert (visually, but also through aural and verbal modes such as music, narration, captions)?

The third question is where our knowledge of semiotic codes and conventions comes into play in explaining how the desired connotations are embedded or '**encoded**' (Hall, [1973] 1980) into the advert. Importantly, our analysis at this stage also has to be alert to irony and humour.

Findings: what does it mean to be 'more dog'?

In this section we return to the three questions posed above, answering the first two relatively briefly and paying most attention to the third 'how' question, to examine the way in which semiotic choices and symbolic attributes contribute to constructing an attractive and humorous advert.

What is the advert selling or promoting in the most basic terms?

The advert is selling the mobile phone network services and products of O2 (phones, tablets, sim cards). The campaign aimed to build a brand identity so that when choosing their next mobile phone network provider, consumers would opt for O2. The campaign also promoted new services such as TU Go, allowing customers to call or text using their mobile phone number via Wi-Fi on a laptop and flexible contracts (Refresh). It is worth noting the context here in terms of other rival mobile provider campaigns – competitors Three had caused a sensation earlier that year with their moonwalking Shetland pony advert using the caption: 'Silly stuff. It matters.' As mobile phones become more associated with web-based activity and data allowances (rather than actual phone calls), the companies appear to have embraced the playfulness and virality of internet culture.

What is the advert selling in symbolic or abstract terms?

The campaign implored viewers to embrace life and the wonders of technology. In this way, it sold an attitude (*carpe diem* – seize the day) and so encouraged engagement, adventure and enthusiasm. It directly contrasted this ideal with indifference, aloofness and disinterest. In embracing a positive attitude, the consumer is assured more fun and joy in life. This was about 'infecting the nation' with positivity – a declared ambition reaching far beyond mobile phone usage.

How are qualities of the symbolic meaning coded into the advert
(visually but also through aural and verbal modes such as
music, narration, captions)?

In answer to this third question, we develop our analysis in stages. The campaign worked across various media, but we focus here on the launch advert. Due to its dominant and striking structure, we start with the slogan 'Be more dog' as a way into the core meaning-making properties of the advert. We then present an overview of the launch advert structure by breaking it down into sound and image, and noting key features. We provide a general description of the compositional features here (lighting, angle, movement, focus, point of view, framing, edits) but *in a selective manner*, based on our analytical interest in the codes and conventions of advertising and especially in the symbolic attributes of the animals depicted. We're interested here in how 'being more dog' is both denoted and connoted in the semiotic choices and relationships. Finally, we focus on the device of 'classification incongruity' employed to generate the principal contradiction and disruptive appeal of the advert.

What does it mean to 'Be more dog'?

The word 'dog' is generally used as a noun, to mean a four-legged canine mammal that barks, but in using 'dog' here as an adjective, something you can be 'more' of, the short phrase is striking due to its apparent erroneousness. The phrase 'be more dog' is also an example of the imperative mood – a grammatical mood that forms a command or request. It directly addresses an implied subject (you) to 'be more dog'. In these simple three words, exhorting you to 'be more dog', the campaign slogan attracts attention and calls into being a particular playful subject. As Judith Williamson writes in her classic study *Decoding Advertisements* (1978: 55), adverts 'appellate' a particular spectator, hailed by the signifying effects: 'it projects out into the space in front of it an imaginary person composed in terms of the relationship between the elements in the ad'. Although the imperative to 'be more dog' is not strictly an element of visual communication, it conjures an incongruous image in the viewer's head through employing 'dog' as a describing word. Working alone, this is still a malleable and polysemic vision, but alongside the other representational devices employed, the advert builds on this ostensibly nonsensical expression.

Outline of the basic structure for the O2 'Be more dog' launch advert

Here we split the visual and verbal elements of the advert to summarize the compositional and narrative formation of the advert in descriptive terms (Table 9.1). Following this, we provide an interpretation suggesting how certain elements work together to create a certain appealing aesthetic, generate humour and associate ideals with the product being sold.

TABLE 9.1 Breakdown of O2 advert into images and sound (vocal and other audio).

Image	Sound [music in square brackets]
Opens with close-up of cat's face, background out of focus	I used to be a cat [no music to start]
Cuts to same cat lounging on a blue sofa, tail swishing [all transitions are 'cuts' rather than fades or dissolves]	Every day the same [soundtrack: slow piano music. *Gymnopédie No.1* by French composer Erik Satie]

Image	Sound [music in square brackets]
Cat lying on deep carpet on the floor as unidentified female hoovers behind him	
The cat staring out of a window, facing away	I'd be aloof till lunch
Cat sits on a worktop as children play in the background [out of focus]	Then coldly indifferent after
Cat sitting on the kitchen hob	
Unidentified female shaking a toy with feathers to amuse the cat while he looks away	To me everything was just 'meh'
Cat sitting on the blue sofa again; he turns to face the camera	Every day the same. Then it hit me . . .
Cat walking in the hall, coming towards the camera	. . . why be so cat?
Cut to point-of-view shot, this time the camera moves towards the cat-flap [new camera movement]	Why not be a bit more . . .
Cut back to cat now running towards the camera (and the cat flap)	. . . Dog!
Cut to outside the blue door and the cat leaps through the flap [slightly slow motion]	[soundtrack changes: Flash! . . . ahhhahh . . . saviour of the universe . . .]
Quick cut to cat landing outside	[soundtrack singing: ahhhahh . . .]
Cat plays with ball on the lawn, now his front paws look more dog-like	
Quick edit: cat circling	[soundtrack: . . . saviour of the universe . . .]
Quick cut to cat ripping up the newspaper	[Soundtrack: guitar riff]
Runs through grass with lens flare effect in low sunshine	[Soundtrack: guitar riff]
Cat/dog hybrid digs a hole in the garden	[Soundtrack: dum-dum-dum, dramatic sound of the piano]
Low-angle shot of the cat as he looks high to the right, with blue sky behind	I mean, look at the world today
Quick cut to running in the garden	It's amazing
Quick cut to running in the street	Running
Quick cut to rolling in the long grass in evening sunshine [lens flare]	Amazing
Cat shown chasing a blue car	Chasing cars. Amazing
Cat carrying a stick in its teeth in a park	Sticks . . .
Cat jumping into a pond at the park [slo mo]	Amazing [music builds]

(Continued)

TABLE 9.1 (Continued)

Image	Sound [music in square brackets]
Cat level angle as unidentified male throws a frisbee as the cat comes towards the camera	*Carpe diem*
Cat leaping off the ground, from left to right	It means . . .
Cat leaping and catching blue frisbee, centre	. . . grab the frisbee
Quick cut to cat swimming in the park pond	
Cat walking past a collie dog in the park, looking until the dog turns around	
Cat runs through the park in low sunlight again	[Music builds again]
Cat figure shakes off water by the pond	[Soundtrack: 'he's for every one of us']
Cat leads group of dogs over a hill in the park, they run towards the camera	Maybe we should all . . .
Cut to slightly closer shot of the cat running towards the camera with other dogs	. . . be a bit more dog
Cat leaning out of a car window, with an apparent smile; lighting is evening sunshine again with lens flare 'Be more dog' appears on screen	New narrator [actor, Sean Bean]: Be more dog
Cut to blue screen, with O2 logo emerging in the centre, and below this 'bemoredog.com' and 'Telefonica'	Start now at bemoredog.com. O2

This summary represents a minute-long advert and we can note the number of different shots (36) within 60 seconds. On average, this represents less than 2 seconds per shot. The edits become quicker as the various dog-like activities are depicted. There is also an alignment between the crescendos in the musical score and the more frenetic activity and editing style. The voice of the cat narrator often bridges the image edits, so that the sound creates a certain effect, in some cases adding humour to the unexpected end to the sentence ('*carpe diem*, it means . . . grab the frisbee'). In most shots there is very little camera movement, so when the camera does move (in a tracking zoom towards the door) it creates a dramatic effect, alongside the main catchphrase ('why not be a bit . . . more dog?') and the introduction of the rock-opera backing track by Queen.

Launched in the summer, the advert uses sunlight and blue skies to connote happiness and the joy of running and swimming in the park on a hot day. Technically speaking, lens flares are considered a mistake because light should not enter the camera in this way, but more recently the effect, whether created as a digital special effect or naturally, is employed widely to capture the feelings associated with sunshine, of warmth and uncontained brightness (Figure 9.1). The colour blue is also used consistently (the frisbee, sofa, car), which both provides a clean, contrasting aesthetic, but also happens to be the company's main colour in its branding. Even within the short advert length, repetition is also used (such as in the word 'amazing' or the shots of running) to hammer home the upbeat and life-affirming message.

Messing with binaries

In this advert campaign, we initially observe how the differences between cats and dogs are highlighted in a humorous manner, with some fun being had, particularly at the expense of 'aloof' cats. Later on, the cat protagonist and narrator becomes 'more dog' and is depicted as a hybrid creature, shaking off water after swimming and catching frisbees but still looking most like a cat due to its facial features. As semiotics teaches us, signs work in relation to other signs, and in this case the humour is created by a confusion of **paradigmatic signs**, to use Saussure's terminology (see Chandler (2007) for a useful introduction). Paradigmatic relationships refer to how meaning is made through differentiation – in other words, the selection of one signifier instead of another, the substitution of comparable but different signs. There is a recognized difference in shaping preferred meaning when choosing one over the other – i.e. choosing a cat instead of a dog.

FIGURE 9.1 Still image from the penultimate shot in O2's 'Be more dog' launch advert. Source: YouTube. The evening sunlight is used to good effect here, evocative of sun-drenched happy times. A small amount of shading is also applied to the lettering providing more sense of depth and movement to a usually flat sans-serif font.

To briefly complete the Saussurean reference here, in this structural understanding of semiotic relationships, the other kind of signifier proposed in this kind of analysis is the **syntagmatic sign**. Syntagmatic relationships concern the ordering, positioning or sequence of signs; how signifiers are positioned in relation to each other, how they are combined together or in spatial sequence. So, whereas analysis of paradigmatic signs imagines the effects of substituting one signifier for another, syntagmatic analysis works through identifying the interplay of the signs' combination in the media text. For a single image, this would mean analysing the co-presence of signs on the page; for moving images, the meaning is generated by the sequence of signs looking at how shots or scenes work to build a narrative form, for example. By adding more signifiers, the meaning becomes more layered and complex.

The O2 advert works by deploying the stereotypical associations of both household pets while simultaneously visually disturbing the binary between cat and dog (in its hybrid character). This central paradox (of a cat becoming more dog) works both with a familiar differentiation and playfully unsettles it.

It combines the appeal of a luxuriously hairy cat with the exuberance of a playful dog. Animals (and especially digitally altered or CGI animated versions) offer advertisers a way to connect with our fascination for looking at animals, our love of furry creatures, while plugging into the symbolic attributes and stereotypical associations of our favourite pets.

Case study conclusion: embracing contradictions

As we hope we've illustrated, analysing adverts does not have to involve obscure technical language and can actually be rather good fun. Choosing an advert campaign that has won plaudits in the industry usually ensures a multi-layered and symbolically rich quality to its visual appeal. The systematic or function-driven approach of semiotic and other visual analyses offer a way into breaking down the elements of representation, finding patterns and connecting intertextual references, but they do not necessarily deal effectively with irony or humour. In some cases, it is fruitful to identify and decode one or two principal devices employed to achieve the desired effect: in this case, a striking use of classification incongruities in both verbal grammar (be more 'dog') and in visual form (a cat who thinks he's a dog). The inventive flair of the advert 'appellates' (Williamson, 1978) an imaginative and knowing spectator. If only our own cats and dogs could speak to us in this way.

But what of ideology? In taking a swerve to analyse an advert we thought captured a visual appeal beyond its immediate purpose, we have neglected some of the more familiar territory of advertising analysis. Taking a more critical stance, the O2 campaign uses the fantasy of talking hybrid animals to sell an idea of a 'good life', in which seizing life's opportunities is associated with using more of their products. From a critical perspective, the advert's artistic merits are deployed to serve the capitalist economic system, specifically mobile phone production, potentially distracting consumers from the reality of the environmental and social consequences of their desire to upgrade their phone every year.

But while it could be argued that the O2 advert aligns purchasing power with happiness, it also overtly celebrates the simple things in life. As Nicholas Holm (2017) argues, the multiple and sometimes contradictory roles of advertising require a sophisticated understanding that resists caricaturing it as either a benign cultural producer of appealing images or a deceptive force for capitalism. Likewise, we should not characterize its producers or audiences in simplistic terms, as creative free minds or passive dupes. Holm suggests a **dialectical approach** that considers competing conceptions at the same time: 'By exploring advertising from the perspective of both capitalism and agency, we are able to better grasp its complicated cultural functions without prematurely reducing it to one role or another' (2017: 207). In this advert, the magic of advertising (Williams, 2005) meets the magical promise of animals, building a brand familiarity associated with an adventurous lifestyle. The clean, sun-drenched world of the cat–dog hybrid offers consumers a fantasy with a humorous twist.

CASE STUDY 2: FILM TRAILERS AS PROMOTIONAL TEXTS: *BLACK PANTHER* (2018)

Due to the expansive nature of visual communication forms we cover in this book, we are unable to cover film analysis in any depth. However, there is a long tradition of film analysis with many excellent guides (Bordwell and Thompson, [2004] 2017; Monaco, 2004). What we can achieve in this space is an analysis of the film trailer. The trailer combines the language of film with the language of promotion. Within this 'language' we incorporate verbal communication (written and spoken), but also music, soundscape and, of course, the visual.

We have chosen the trailer for *Black Panther* (2018), which lasts for just over two minutes and is available on Marvel Studios' YouTube channel (www.youtube.com/watch?v=xjDjIWPwcPU). This film trailer won the most plaudits at the Golden Trailer awards, winning best 'action' film trailer and also the overall top award, 'best of show'. The film itself has entered the top ten 'all time' highest grossing films (along with three other *Avengers* franchise films at the time of writing). The trailer, known as 'Crown', was produced by the Create Advertising Group for Marvel Studios, a subsidiary of Walt Disney Studios. The actor Chadwick Boseman had already starred as the Marvel comic book superhero in *Captain America: Civil War* (2016), but the later character-led *Black Panther* attracted critical acclaim and much commentary due to its near all-black cast. The film was deemed radical in some quarters, depicting a futuristic African nation that had never been colonized, female warrior guards and a 'baddie' nemesis who teaches the titular superhero a moral lesson.

Behind the cameras, *Black Panther* also had a black director, Ryan Coogler, who co-wrote the screenplay with black screenwriter, Joe Robert Cole. At a time in which the #MeToo movement has profoundly shaken Hollywood in terms of reassessing how women are treated and valued in the industry, a wider discussion on inclusion and diversity continues to bring attention to how stars can work to ensure gender and racial equality in hiring on movie sets, writing an 'inclusion rider' into their contracts. The worldwide commercial success of *Black Panther* has brought hope that it could be a 'game changer' in challenging the industry bias that films made by and starring people of colour do not 'travel' well globally, and in proving that 'diverse audiences want diverse stories' (Levin, 2018).

Research questions and approach

Film scholar Lisa Kernan provides one of the most comprehensive studies of Hollywood film trailers in her book *Coming Attractions: Reading American Movie Trailers* (2004). It is the trailers' 'unique' rhetorical structures and hybrid nature as both narrative and promotional texts which in turn make them so revealing about the 'ideological and cultural conditions of, or constraints within, Hollywood cinematic narrative itself in specific historical moments' (2004: 7). These 'anticipatory' texts transform 'the codes of narrative fiction into the codes of promotional rhetoric' and so construct a new '*trailer* logic' where the film's 'shiniest wares' are displayed (2004: 10).

In her analysis of Hollywood trailer rhetoric, Kernan focuses on three appeals, which often work together in a complementary relationship: **genre, story and stardom**. Genre is a complex term to unpick here, but regular film viewers are able to imagine generic conventions of, say, a romantic comedy, a crime thriller or horror film, and would be able to differentiate between them at a glance. Through in-depth analysis, the features of each **genre** can be examined to see how films are sold to audiences through replication of certain features, and in commercial terms this serves to 'remind audiences of their own attachment to this kind of repetitive and ritualized spectatorship' (2004: 49).

One common criticism of trailers is that they give away too much of the story. In Kernan's model, '**story rhetoric**' concerns how much of the story arc is revealed in the trailer and how much is withheld, hopefully generating enough interest while maintaining suspense. Voice-over narration or selected fragments of dialogue introduce audiences to the characters, often sleekly providing cues about the heroes and villains. This characterization aspect of story rhetoric connects to the final appeal – **stardom**. Unlike genre and story, the rhetoric of stardom 'possesses an indexical relationship to the social world' (2004: 63): the appeal here is not related to the character in the film-story, but works with prior knowledge of the stars' acting roles and with their celebrity persona. The rhetoric of stardom assumes and even generates excitement about seeing the next film of a certain star, their allure operating beyond the confines of the film itself. As Kernan argues, each of the three appeals – genre, story, stardom – makes assumptions about audience interests and so provides 'textual evidence' of the broader ideological assumptions of the industry, and allows the researcher to observe shifting patterns over time.

In this case study we use a (much simplified) version of Kernan's triadic trailer rhetoric, and break this down further by analysing 'voice-over narration, sound and sound overlapping, music, graphics and most importantly, editing, or montage' (2004: 10). While our focus here is on the visual mode of communication, we recognize that meanings are made in intersection with other modes. Indeed, we would suggest that trailers apply a particular intensity of semiotic or rhetorical modes, frequently working together towards some kind of spectacular crescendo within the trailer (rapid editing, rising music, resolute voice-over, film title, date of release). Therefore, the case study's overarching research question is:

- What do the interlinked rhetorical appeals of genre, story and stardom reveal about the promotional strategy for the film and imagined audience desires?

Findings: celebrating blackness in the superhero genre

Summary of the *Black Panther* (2018) trailer

The hip-hop music soundtrack leads the way, with an initial black screen fading in to aerial shots of dramatic waterfalls and open landscapes, through which spaceships fly. The disembodied opening narration sets the scene without linking the voice visually to any character: 'I have seen gods fly, I have seen men build weapons that I couldn't even imagine, I've seen aliens drop from the sky, but I have never seen anything like this.' The repetitive refrain on *seeing* fantastical spectacles has a dual function; it is presented as both diegetic dialogue (within the anticipated film's world), but also as a way to excite expectations about the coming film ('never seen anything like this'). As the film cuts to an inside scene, CIA agent Everett Ross asks, 'How much more are you hiding?', to which T'Challa simply turns his head to face him remaining tight-lipped. The sharp cut to the familiar red and black branding of Marvel Studios is timed to coincide with the music building and the soundtrack singer exhorting, 'Let's go'.

The next sequence of shots suggests arrival by spacecraft to this beautiful land (the fictional Wakanda) both densely forested and quirkily urban in its combination of Afrofuturistic houses and skyscrapers, as Okoye announces 'We are home'. Each shot is only around a second long, with a mix of fades and cuts that introduce Black Panther's female connections in various roles. The frequent dissolves to a black screen not only work in time with the musical soundtrack, but suggest that each revealed scene holds new delights to savour. We also see the requisite superhero trope where the costume is revealed and he transforms from T'Challa to Black Panther, complete with futuristic weaponry (Figure 9.2).

The next segment follows a quick title break displaying 'in 2018', before introducing T'Challa's nemesis, Erik Stevens (who becomes Killmonger). His character is established in a scene where he breaks a prisoner out of custody using explosives. This middle section frenetically depicts fast-moving action in various locations, interspersed with the credits at regular intervals: 'Hero', 'Legend', 'King', each appearing in gold lettering on a black screen. The short section that follows the 'Legend' title page is in stark contrast to the others in the sequence, depicting a purple hued spiritual world where T'Challa approaches an acacia tree in white robes under aurora borealis skies (Figure 9.3). Acacias, with their

FIGURE 9.2 Black Panther/T'Challa surrounded by his female warrior guards, mixing futuristic technology and African tribal costumes.

distinctive wide canopy, are often used as a symbol for sub-Saharan Africa, with many travel magazines and novel covers opting for a silhouetted tree in the setting sun. In this very short sequence, the visual shorthand for African spirituality is strikingly slotted between shots of high-tech weaponry and car chases.

FIGURE 9.3 The purple colours and spiritual world contrast sharply with other scenes in the montage.

The following section sets up the battle between the two opposing forces, as clashes are depicted in a range of location settings, with a particular focus on the Asian cityscape towards the end of the trailer. Throughout, the trailer utilizes an array of unusual camera angles, including tracking rotations from upside-down to sideways, canted angles from below firing weapons, slow-motion rotating aerial shots of overturning cars and a jumping Black Panther, in addition to other computer-generated breathtaking effects. Sound effects are used to emphasize the speed and impact of various weapons and vehicles, depicting futuristic technology as well as hand-to-hand combat. The pacing and camera movement are so unrelenting that any slight pause in editing or stability in the *mise-en-scène* suggests a moment requiring the viewer's special attention (as with the acacia tree scene or the slow-motion car flip). The final fade to black brings up the title 'Black Panther' followed by the date of release.

Before turning to our observations on genre, story and stardom, we want to comment on the opening hip-hop soundtrack that continues throughout the trailer. The spoken-word track 'The revolution will not be televised' (1971) by Gil Scott Heron, is here meshed with 'BagBak' (2017) by Vince Staples. It is this aural element (rather than the visual aspects) that signals a radical intent. The 1971 song is associated with the Black Panther movement of the 1960s and 1970s – a name match that Marvel had previously distanced itself from. In sampling this track about revolution in a mash-up with a song by a contemporary hip-hop artist who is also known for lyrics that address police brutality, injustice and racial inequality, the trailer suggests political awareness and even consciousness raising as part of its appeal. Alternatively, another perspective might bemoan how a song about turning off the television to participate in the fight against injustice has been subsumed into the promotion for a superhero film.

Genre rhetoric

As part of the Marvel Studios *Avengers* franchise, the film trailer follows the recognizable style of an action film and more specifically one inspired by a comic book universe. In line with this genre, the trailer is packed with physical action (violence, chases) and special effects, rapidly cut together across diverse global and sometimes fictionalized locations. Short bursts of dialogue suggest the magnitude of the challenges faced by the main characters, but also hint at the humour shared. While spectacular violence features heavily, depictions of sexual desire are absent. There is no separate epic trailer voice-over, but we do hear and then see T'Challa declare, 'What happens now determines what happens to the rest of the world'. In *Avengers* films, the stakes are always high.

But what about the ways in which it challenges or innovates? The African motifs give this film an imaginative aesthetic linked to Afrofuturism, an artistic sensibility or cultural field that aims to break down binaries between the mythical and scientific, of past and future, and in doing so, challenge myths of blackness inscribed by white-dominated culture. 'By stepping outside of the white liberal tradition and rewriting blackness in all its complexity, Afrofuturism offers a novel form of revolution that is rooted in a long tradition of black opposition' (Rollefson, 2008: 105). For an excellent introduction to the influences of such artists across science fiction, avant-garde hip-hop and graphic novels, see Ytasha Womack's book *Afrofuturism: The World of Black Sci-Fi and Fantasy Culture* (2013). In relation to our case study, the attention paid to the textures and materials of tribal costumes, combined with futuristic science and technology in a hidden utopian land, have seen Black Panther hailed as a new empowering symbol of an Afrofuturistic renaissance (Fitzpatrick, 2018).

Story rhetoric

The film trailer is confident enough in the popularity of the series not to include any explicit intertextual references to preceding films beyond the Marvel Studios branding. For fans, the arrival 'home' follows on chronologically from the assassination of T'Challa's father in *Captain America: Civil War* (2016), and so signals his ascension to the throne in Wakanda. But without this intertextual knowledge, the trailer follows a classic narrative and simplifies the main story line into a battle between Black Panther and Killmonger. Indeed, in story terms, the later car chase scenes in an Asian setting (and filmed in South Korea) appear bolted on without any clear narrative consistency. Marketing imperatives are arguably hard to disentangle from the film's story world, and the choice of South Korea for the other key filming location provides both a spectacular setting for a high-octane chase through colourful markets and landmarks, but also hints at the importance of the Asian film economy.

As is often the case with action film trailers, genre appeals override plot details or story rhetoric, and beyond the differentiation between Black Panther's identity and his enemies, little of the actual story arc is revealed. While causality is left undeveloped, the sense of destiny is apparent. Additionally, the story world of Wakanda is lovingly created and even the small glimpses offered here of the architecture, costumes, technology and mythology suggest a celebration of Africa smuggled into an overly familiar Hollywood genre (see Figure 9.4).

Stardom rhetoric

One irony of the trailer is that CIA agent Everett K. Ross (played by Martin Freeman), one of the few white characters, gets a line. Freeman is known for his comic timing and therefore his presence could provide a promise of humour due to his other known roles. Freeman's particular quality of *relationality* is being deployed here – especially perhaps for white audiences (in terms of industry assumptions about identifying with sameness, not necessarily a comment on their actual responses).

FIGURE 9.4 Skyscrapers and colourful markets collide in Wakanda.

Generally speaking, however, stardom in one sense is the weakest logic here. No stars' names appear as part of the promotional strategy within the trailer, and while there are a number of recognizable faces on screen, the hyperbole comes in the form of the spectacular production design and computer-generated effects. If, following Kernan, trailers provide 'textual evidence' of the broader ideological assumptions of the industry and their evaluation of audience interests and desires, the identity of the star is not the principal appeal.

In her own conclusion, Kernan notes the way in which categories have become more fluid with stars becoming franchises: 'Stars "become" genres, as these formerly unified textual categories of trailer rhetoric become increasingly synergistically interwoven with the high concept promotional environment' (2004: 216). In our case, the franchise becomes the star. The intertextual associations evoked here are with the other *Avengers* films. The 'promotional environment' now includes the easy availability of further information and promotional material about the film online, which might well play a part in the omission of star appeal in the trailer. We suggest that the stars play second fiddle to the 'Marvel Cinematic Universe' film world, which appears to subsume genre, story and stardom appeals in its apparently unstoppable Hollywood ascendancy in recent years.

Case study conclusion: *Avengers* meet Afrofuturism

Kernan's trailer logic of genre, story and stardom provides a framework for thinking about how the trailer is structured, how certain elements in this recipe are highlighted over others, and the degree of intertextual and meta-textual references relied upon in the construction of its appeal. It can also help us to recognize what makes each film trailer different (or indeed similar) once compared with others. In the global film market, it is also worth examining whether trailers for the same film are adapted for local markets – what might this reveal about assumed ideological or cultural differences in the global entertainment industry?

It took 50 years for the original Marvel comic book character of Black Panther to make it into cinemas, but the global critical and commercial responses have propelled the film into Hollywood history as a 'game changer'. In reality, it is likely to be part of a broader cultural shift already underway, but the celebratory tone is arguably a response to its stunning visual style of Afrofuturism (Womack, 2013), as well as the visibility of black actors and the creative direction of Ryan Coogler behind the screen. Analysing trailers could be said to focus attention on the marketing of films rather than their cultural significance, but following Lisa Kernan cited above, we agree that trailers provide 'textual evidence' of the broader ideological assumptions of the industry. In this case, what is noticeable is the visionary rendering of a sublime African landscape among the more usual hi-tech weaponry and impossible action sequences associated with the *Avengers* franchise.

We cannot yet know its lasting influence, but many black writers have already hinted at the empowering nature of the film for young diasporic African audiences who finally get to see role models who look like them, whether as young female scientists or superhero warriors. As Olive Pometsey (2018) sums up in *GQ* magazine: 'It's one thing to see a film where Hollywood's black talent get a chance to shine, but it's quite another to watch them create a world where their race isn't weighed down by the oppression and prejudice that they face in real life.'

CHAPTER CONCLUSION: THE POETICS OF PROMOTIONAL CULTURE

In this chapter, we have focused on the allure of promotional imagery. By concentrating on the format and structure of adverts and trailers we have highlighted their 'poetics' over their 'politics' to some degree (Hall, 1997). Indeed, we have selected examples which we thought would be fun and interesting to analyse. These are introductory case studies, but they hint at potential areas for expansion into more critical areas of study. The emerging field of animal studies could help us to understand how and why we integrate other-than-human animals into cultural artefacts, and how this changes over time and with emergent media technologies. Our second case study identified continuities in the structural features of the action film trailer, but also suggested the integration of an Afrofuturist aesthetic that signalled a radical and potentially empowering intent.

Early on in this chapter we set out Nicholas Holm's (2017) five reasons to study advertising: advertising is everywhere; advertising is weird; advertising is where the money is; advertising is beautiful, inspiring and entertaining; advertising is political. Although admittedly limited to a textual analysis approach here, we hope to have demonstrated Holm's reasoning by closely examining some of the techniques and mechanisms of advertising and promotional culture.

FURTHER READING

The updated edition of Marita Sturken and Lisa Cartwright's *Practices of Looking* (2017) contains further theoretical background and examples of advertising analysis. David Hamel's (2012) article on 'teaching with trailers' demonstrates how Bordwell and Thompson's ([2004] 2017) four-step process of 'Analyzing Film Style' can be applied to trailers in order to develop critical analytical skills.

10 VISUALIZING LIFESTYLES AS COMMODITIES

It is not only products and services that are sold in the media, but also guidelines on how to make choices in matters of health, fitness, nutrition, fashion, leisure, home decor, cuisine, and even work and family. In this chapter, we consider lifestyles as commodities in their own right and visual imagery as central to promoting ways of being and living that are largely rooted in consumerism. In sum, this chapter will:

- introduce research on lifestyle media and the role of visual communication in defining lifestyles as commodities;
- explore the increasing significance of lifestyle politics in consumer culture and its relationship with corporate media and commercial imagery;
- show how a countercultural lifestyle aesthetic is communicated in corporate media through a combination of different visual cues, with a focus on the official website of Vice Media (Case study 1);
- examine some of the ways in which a feminist perspective is communicated through lifestyle imagery, particularly in stock photographs from the Lean In Collection by Getty Images (Case study 2).

The chapter's case studies focus specifically on corporate media and **commercial imagery** that 'sell' lifestyles based on political values traditionally outside of the mainstream. In addition to highlighting this growing aspect of media culture, this focus allows us to further demonstrate the key role of the visual in promoting something as immaterial and complex as a political outlook as a way of life. Before we delve into our case studies, however, we need to spend some time reflecting on the notion of lifestyle, both in terms of the theories underlying its definitions and its relationship with contemporary media.

LIFESTYLES, CONSUMER CULTURE AND THE MEDIA

The notion of lifestyle pertains to the ways in which individuals act routinely in their everyday lives, particularly as a reflection of their cultural values, economic status and social aspirations. In traditional societies, lifestyles are closely linked with the ways in which given groups or communities organize their shared patterns of behaviour, often in relation to their position in broader society as a collective (Weber, 1978). Meanwhile, the rise of consumer culture has brought freedom of choice and self-expression to the fore (Featherstone, 1987), thus contributing to equating the notion of lifestyle with **individuality**. This ability to craft a coherent and unique narrative about oneself by choosing from the endless amount of options that are seemingly available to all in consumer societies extends well beyond decisions in matters of consumption alone. It also encompasses deliberate choices about the way one speaks and acts, the career and relationships one pursues, and the ways in which one integrates political, ethical and spiritual outlooks into daily practices (Portwood-Stacer, 2013). It is in this sense that lifestyle becomes what Giddens calls a 'reflexive project of the self' (1991: 5).

Contemporary media promote ways of being and living that are linked to individuals' ability to make informed decisions on how to present oneself, how to interact with others and, ultimately, also what to consume. 'Lifestyle experts' offering advice on personal taste and appearance are widespread in popular media. Many of us are now familiar with makeover shows on television, where 'struggling' individuals and 'deserving' families are given glamorous looks and brand-new physiques or shiny new homes, respectively. Celebrity chefs like Jamie Oliver and media personalities like Oprah Winfrey provide guidance on everything from food and finances to health and self-improvement across their branded television shows, websites and print publications. And while women's magazines still feature advice columns and articles that deliver lifestyle guidelines in an informal yet informative tone, the rise of bloggers and Instagram celebrities has further extended the reach of **lifestyle expertise** to 'ordinary' people (Bell and Hollows, 2005). Print magazines, television and the internet are all major outlets for the promotion of everyday tastes that are firmly rooted in consumer culture (Lewis, 2008). Across media outlets and genres, visual communication is key to defining the distinctive 'look' of particular lifestyles as goods that are just as intangible as they are pervasive in contemporary promotional culture. In the next section of this chapter, we introduce key research on the visual dimensions of lifestyle media. In doing so, we also offer ideas for the application of the chapter's analytical tools to other potential case studies in this area.

AESTHETICS, LIFESTYLES AND CELEBRITY

Aesthetics and lifestyles go hand in hand. In order to express our personal tastes, values and attitudes, we most often rely on cultural signifiers that make our identities as individuals visible to others. The choices that we make with regard to the way we dress and style our hair may indicate that we are, for example, both elegant and high powered while also being fashion-forward, sporty and creative. In combining a crisp white button shirt and a well-cut, expensive blazer with stylish jeans, athletic shoes and a spiky or asymmetrical haircut, we may very well communicate all of these meanings about ourselves, and therefore also promote a particular narrative about our lifestyle identity.

These are what Machin and van Leeuwen (2007) define as 'composites of connotation', or combinations of signs that are meaningful to the majority of people in a given society but which are not based on predefined roles tied to social categories like gender, class, nationality or race. These lifestyle-related visual signifiers are most often picked up from the media, and especially from **celebrity culture**. Machin and van Leeuwen use David Beckham as a key example in this regard. The famous footballer's identity, as it was portrayed by the media in the early 2000s, combined 'aspects of the rebel and the model citizen' (Machin and van Leeuwen, 2007: 51). The media offered innumerable images of Beckham's personal style, which combined transgressive tattoos and casual trainers with sophisticated designer clothes and 'metrosexual' hairstyles. In doing so, they also provided ample coverage of Beckham's everyday life, showing him not only as a hot-headed sportsman and a wealthy West Londoner, but also as a loving father and an activist campaigning against racism and in support of disadvantaged children.

By foregrounding celebrities' personal attributes and setting them against the backdrop of their everyday surroundings, together with their professional and charitable endeavours, this kind of media imagery also suggests that there is a correlation between these individuals' qualities as citizens and their physical attractiveness and glamour. Through his careful self-styling and media portrayals, David Beckham set new standards for men's lifestyle culture. As a lifestyle icon, Beckham was a precursor to contemporary celebrities' love affair with social media platforms like Twitter and Instagram, to the extent that media personalities like the Kardashians have become famous for being famous – that is, for having thousands of social media followers keen on witnessing their luxurious lifestyles. The world of celebrity, then, is deeply intertwined with lifestyle media as these are individuals who quite literally embody the aspirational ways of life that are key to consumer culture.

SEEING LIFESTYLES IN THE MEDIA

Ultimately, lifestyle media are in the business of providing audiences with toolkits for the acquisition of knowledge and skills that will make them distinctive as individuals, while also directing them towards the 'right' choices in matters of taste and behaviour. Hence, both the contents and styles of visual images are fundamental in defining the particular ways of being and living that are promoted by different media texts and outlets. The lifestyles that are promoted and visualized in the media are by no means unrelated to traditional social structures and, in fact, are deeply connected with implied assumptions and aspirations about class status and gender norms in particular. Reality-based television programmes often purport to enable their participants to transform into a radically better version of themselves regardless of their social background or bodily appearance. Makeover shows in particular rely on an 'emotional aesthetic' (Lewis, 2008) made of visual styles that lay claim to the real combined with visual content that offers evidence of the lifestyle journey undertaken to leave behind an old, undesirable self in favour of a new, healthier or more attractive identity. From a visual standpoint, **makeover television** relies heavily on stylistic choices that foreground the reality of the journey undertaken by the protagonists, while also dramatizing their trajectory – for example, through fly-on-the-wall footage that shows the protagonists' pitfalls in relation to their health or appearance.

The main protagonists of these television shows are usually **ordinary people** whose everyday lives are far from being glamorous, but who are being given the opportunity to be educated by experts like the five presenters from *Queer Eye for the Straight Guy*, each with his own specialist knowledge in areas ranging from personal grooming to culture. Makeover shows also enable participants to access the material resources that are necessary for them improve themselves and their lives – for example, expensive clothing, makeup or even surgery. Hence, documentary styles and before and after imagery are used both to factually evidence the protagonists' transformation, and to add an emotion-laden and often sensationalist tone to each twist and turn in the narrative.

Just like makeover television shows, **lifestyle magazines** provide their audiences with guidelines for self-improvement through changes in one's diet, exercise, clothing and attitude. However, magazines tend to adopt an 'informational aesthetic' (Lewis, 2008) that foregrounds in-depth information that can be accessed at once (e.g. through weekly meal planners or targeted workout plans) over the development of a dramatic narrative over time. Both print magazines and their online versions draw from a combination of articles offering specialized knowledge, first-person success stories and advice columns. In their analysis of *Cosmopolitan*, Machin and van Leeuwen (2007) highlight that the increasingly global advice column format is based on the 'problem-solution' formula, which is typically used to offer 'solutions' to 'problems' regarding anything from romantic relationships to workplace politics. And just like reality television, lifestyle magazines are usually based on global formats that are then adapted for local, culturally specific content conforming to the 'norms' of any given society.

Needless to say, images are key to lifestyle magazines. In their work on the Chinese women's magazine *Rayli*, Chen and Machin (2014) trace the evolution of imagery in the magazine's history over 17 years, noting that the photographic images used across articles have become less naturalistic and more conceptual, where women are portrayed in idealized, decontextualized settings and through the aid of simple props rather than by showing them as performing particular actions in a specific environment. This is an aesthetic that is typical of stock imagery.

Stock photographs (and now also stock footage) are the 'wallpaper' of consumer culture (Frosh, 2003), insofar as the bulk of images we encounter in our everyday life are ready-to-use, generic images that come from global image banks such as Getty Images and Shutterstock. **Stock imagery** is most often associated with a very limited aesthetic repertoire of bland and clichéd images with little or no contextual detail. However, the aesthetics of 'genericity' that are typical of stock images (Machin, 2004) have become increasingly complex and, as we will see in our second case study, also deeply entrenched with current visual trends in lifestyle media.

Traditional advertising is still a major vehicle for the promotion of aspirational lifestyles, and often print ads use some of the same visual conventions of the stock images found in lifestyle magazines and the lifestyle sections of major newspapers. In their work on **luxury tourism advertising**, Thurlow and Jaworski (2010) explain that across a broad sample of print ads for luxury resorts, the privileged class status of their implied customers is communicated through a series of visual and linguistic references to the super-elite lifestyles of those very few people on Earth who can afford to own an island or travel first-class internationally. Therefore, visual signifiers evoking silence and exclusivity like solitary beaches, intimate settings for two, and impeccable service without interactions with servers are abundant across these advertisements. Not surprisingly, these ads are not actually targeted at the super elites (who, after all, may very well own their own tropical paradises), but rather those of us who may be able to afford a once-in-a-lifetime luxury holiday and thus need to be seduced into using their savings in this way.

Lifestyle media are by no means limited to makeover television, women's magazines or advertising, but span across media genres. If anything, and as others have argued, we are witnessing an increasing **'lifestylization' of all media** – for example, through the growing space that news media dedicate to lifestyle content, the presence of ordinary people as 'experts' on television and the pervasiveness of user-generated content on the internet. With the rise of Instagram, blogs and lifestyle websites, both ordinary people and celebrities are now carving out careers out of their ability to dispense advice on the most disparate topics. From 'mommy bloggers' to Gwyneth Paltrow's 'clean living' website Goop, the internet is quite literally exploding with lifestyle advice. Blogs and websites lend themselves particularly well to a mixing of visual resources. As we will see in Case study 1, a careful orchestration of web design elements like layout, typography and colour is fundamental in the creation of a coherent lifestyle identity for a blog or website.

Along the same lines, lifestyle politics have become key to consumer culture and, as a consequence, also lifestyle media. Portwood-Stacer (2013) defines lifestyle politics as the everyday ecology of practices and choices that are used by individuals, particularly feminists and anarchists, to express their political convictions. Niche marketing and the media's co-optation of subcultural styles together with 'sustainable' or 'ethical' lifestyles are nothing new, but the internet has contributed to the popularity of genres like femvertising and hashtag activism, for example. Because of the growing prominence of 'alternative' lifestyles in contemporary media culture, our case studies focus specifically on the appropriation of feminist and countercultural values by corporate media in an attempt to 'sell' lifestyle politics, rather than consumer lifestyles alone.

The first case study examines the homepage of Vice Media's official website, Vice.com, to understand how this media company's signature countercultural aesthetic is communicated through the rich mixing of writing, images, layout and other modes of communication that can be found in digital texts. The second case study explores how a certain version of feminism based on notions such as empowerment and authenticity is promoted, visually, in stock photographs from the Lean In Collection by Getty Images.

CASE STUDY 1: VICE MEDIA'S COUNTERCULTURAL LIFESTYLE IDENTITY BETWEEN STYLE AND SUBSTANCE

Often dubbed as the 'Hipsters' Bible', Vice Media is a Canadian–American media company that started out in 1994 as a small counterculture magazine named *The Voice of Montreal*. Renamed *Vice* in 1996, the magazine quickly became popular among Millennials thanks to its provocative content and rebellious style. In its early days, *Vice* was known for its punk edginess. With the magazine's progressive growth into an internationally known brand, the magazine's three original founders expanded Vice's business model to several other areas of media production and distribution.

Since the early 2000s, Vice Media has become a digital media giant with multiple channels – including Vice News, the music channel Noisey and the food channel Munchies, just to name a few – together with a film and television production division, and a record label (Martinson, 2015). With headquarters in fashionable areas of New York and London such as Williamsburg and Shoreditch, Vice Media continues to promote itself as an alternative to mainstream media and, in doing so, also caters specifically to young people and trend-conscious adults. Ultimately, Vice Media's countercultural appeal relies heavily on claims of transparency, transgression and originality, especially in relation to the lifestyle identities of the media brand's consumers. Therefore, it becomes all the more interesting to understand how the Vice aesthetic is designed to appear in line with countercultural lifestyles.

Research questions and approach

In this case study, we ask three main questions:

- What are the key visual and multimodal resources that set apart the Vice aesthetic?
- How do these semiotic resources contribute to promoting a countercultural lifestyle identity as Vice Media's key selling point?

- What are the critical implications of Vice Media's 'style' in relation to its 'substance', in particular in relation to the media company's economics and ownership structure?

Here, we propose an approach rooted in multimodal critical discourse analysis as we aim to understand how the different **semiotic resources** that are used to design Vice.com contribute to defining a coherent identity for Vice Media as an alternative media outlet in spite of its corporate ownership and controversial views. It is also for this reason that we combine our multimodal analysis of Vice.com with a political economy perspective to problematize the relationship between Vice Media's 'style' and 'substance'. What this means is that an appraisal of Vice Media's economics and business strategies is key to an understanding of the politics and power relations underlying the media outlet's lifestyle ideologies.

Findings: Vice.com's multimodal aesthetics of transparency, transgression and originality

In her social semiotic work on the aesthetics of blogs and web pages, Adami (2015) demonstrates the importance of understanding how a combination of different modes of communication beyond writing is actively used by web designers to produce coherent narratives. This multimodal framework includes an assessment of layout, font, colour, images and writing, together with considerations about interactivity as a key affordance of web-based digital texts. This analytical approach is not aimed at unveiling the intentions of a 'real' author in communicating to a 'real' audience, but rather key 'designed identities as projected by the text' (Elisabetta Adami, 2015: 49). In other words, we often encounter and engage with web pages – and, more broadly, a variety of texts – that may be designed to look amateur or professional, feminine or masculine, sophisticated or rustic and, for example, mainstream or countercultural regardless of whether or not their 'real' producers and consumers fit into any of these social categories.

Therefore, in this case study we examine Vice.com's homepage in relation to the meaning potentials conveyed by **layout, font, colour** and **images**. We leave both writing and interactivity out of our analysis, as these are modes that are not strictly visual. We also add considerations about Vice's logo, which can be considered as a multimodal text in its own right insofar as it is made of typography, colour and language. As we explain in more detail in Chapter 11, analysing logos is especially important because they are the main 'face' of a brand (Lury, 2004). Overall, we aim to understand how the **designed identity** of Vice.com's homepage may promote a sense of identification with a countercultural lifestyle among the website's users.

Logo

Starting from the *logo*, then, we can begin to examine the aesthetics of Vice.com in relation to the lifestyle identity that is evoked by its design. This particular logo is designed as a logotype, because it contains the name of the brand. The word 'Vice' is rendered as a hand-style graffiti made of chunky lettering and a bold contour (Figure 10.1). Hand styles are the unique forms of lettering that graffiti artists, or 'writers', develop to leave their own distinctive 'tag'. They demonstrate a particular writer's individual skills, both in terms of technical competence and creative talent (Ross, 2016). Hand styles are also seen as the root of all graffiti. It is in this sense that, through its logo, Vice can be associated with notions of uniqueness and rawness. While the relative irregularity of the logo's lettering may be linked to meanings like '"authentic", "individual", "personal", and so on' (Johannessen and van Leeuwen, 2018: 191), the logo's continuous, bold contour offers a unifying framework that makes the brand look cohesive, solid and self-contained. In addition, the white space of the lettering itself can become 'transparent' and be adapted to a variety of media surfaces like, for example, the cover of *Vice* magazine. In merging the logo's positive space with the media outlet on which it is placed, the Vice brand is made to look 'close' to the media content that it represents, and therefore also more truthful and genuine.

Overall, the Vice logo balances meaning potentials of authenticity, originality and credibility. The fact that it looks as if it was traced by hand in a graffiti style also adds **experiential metaphors** related to ideas of transgression and rebellion against the norms of dominant culture and mainstream media institutions. As van Leeuwen (2006) explains, the notion of experiential metaphor originates from Lakoff and Johnson's (1980) groundbreaking work on the important role that linguistic metaphors play in shaping our perceptions and actions in everyday life. When applied to an analysis of lettering and typography, it refers to the idea that a particular signifier may have 'a meaning potential that derives from our physical experience of it, from what it is we do when we articulate it, and from our ability to extend our practical, physical experience metaphorically, to turn action into knowledge' (van Leeuwen, 2006: 146–7). Ultimately, our embodied memories of encounters with graffiti and an understanding of what it takes to trace them on city walls contributes to the Vice logo's countercultural aesthetic.

Layout

In addition to being a multimodal text in its own right, the Vice logo is also an integral part of the Vice Media website's layout. The logo is placed in the middle of a horizontal banner at the top of the page, which includes no other information but a drop-down menu symbol on the far left and the text 'VICE CHANNELS' followed by a drop-down

FIGURE 10.1 Vice logo.

arrow on the far right. Information about the several sections of the website together with links to social media and contact information as well as the main eight Vice channels (e.g. the food channel Munchies and the music channel Noisey) is relegated to these drop-down menus, thus contributing to the website's stripped-down aesthetics. In her analytical framework, Adami (2015) makes a distinction between **orientation and framing** when it comes to web page layout. Orientation refers to the way in which text is arranged in the different sections that make up the space of any given web page. It can be vertical or horizontal, and web pages often feature a combination of the two. Framing, instead, refers to the way in which the different sections of a web page are separated or connected – for example, through more or less regular spacing or the use of lines to demarcate internal borders.

The layout of Vice.com's homepage is strikingly horizontal in terms of orientation, as content is mostly organized in rows rather than columns. For example, the main feature occupies the entire layout of the homepage before one scrolls down the page to see additional content, and all secondary features are laid out in rows divided into two columns that split the homepage exactly in half, with the left column entirely filled by an image in landscape format and the right column occupied by the headline, summary and journalist's name (Figure 10.2). The Vice.com's homepage layout is also in part modular, as latest stories from different categories (e.g. 'LGBTQ', 'FILM', 'WAR') are arranged as 'newsbites' across three rows and four columns. As Adami (2015) suggests, a largely horizontal and

FIGURE 10.2 Horizontal orientation layout in Vice.com's homepage.

modular layout evokes a more contemporary rather than traditional approach to design, both in online and print media. In terms of framing, it is mainly white space and light grey lines or bands that are used to separate elements within the page. Spacing between different elements and sections is wide and regular, both in terms of text alignment and distance between different elements. The contemporary, regular and spaced layout of this homepage can be associated with minimalism, thus infusing the Vice brand with a modern and cutting-edge feel.

Font

The main font used in the Vice.com website is Helvetica, which is not only the most widely used typeface in advertising and consumer culture, but also one that has become increasingly associated with a 'hipster' aesthetic, thanks to brands like American Apparel and Urban Outfitters together with the now iconic signage of the New York Subway system (Rose, 2014). Helvetica is as ubiquitous in corporate branding as it is in DIY culture, as its sans-serif, modernist design makes it especially 'clean' and versatile (Garfield, 2010). For a lifestyle media outlet like Vice, Helvetica may be an even more obvious choice than other sans-serif fonts due to the fact that it is one of the most widely used typefaces in urban space, or as Norwegian designer Lars Müller (2013) puts it, Helvetica is 'the perfume of the city'.

In Vice.com's homepage, Helvetica is used for headlines, summaries and the main text of stories used in a range of sizes and weights. In an article about their work on the visual identity of Vice's television channel Viceland, New York branding agency Gretel stated that 'it was asked to create an identity that could express the tone and personality of each of Vice's shows without overshadowing the content' (Steven, 2016). The choice of Helvetica Bold for Viceland's 'unbranded' branding was related not only to its simplicity, a quality that other sans-serif fonts share with Helvetica, but also to its 'sense of modernism' and the fact that it felt both 'pure ad 'contemporary' (Steven, 2016). Perhaps precisely because of this tension between Helvetica's simplicity and edginess, this font choice contributes to the overall Vice aesthetic by conferring meaning potentials of transparency and transgression to the website's homepage.

Images and colour

The photographs used across the homepage tend to be high in modality, insofar as they largely conform to key conventions of news images, both in matters of newsworthy content and naturalistic style (Figure 10.3). The overall colour scheme – light grey, white and black – is quite minimalistic, thus contributing to the broader aesthetic of simplicity adopted in the website design. Together, images and colour contribute to some of the same visual claims of transparency that we discussed earlier. Vice.com's colour scheme adds to its cutting-edge appeal. The homepage also mixes different styles of photography, with

high-modality images being prevalent, but with a tendency to use also both grainy and highly saturated photographs in the 'hipster-sleaze' style of Terry Richardson, the highly controversial photographer who used to be one of Vice's key contributors before being accused of sexually preying on young models. There is a fairly stark contrast between these kinds of images and the website's colour scheme, which further solidifies Vice.com's countercultural aesthetic as a combination of visual claims of transparency, transgression and originality.

FIGURE 10.3   'Hipster' photography in Vice.com's main feature.

Case study conclusion: selling style over substance?

Through visual and multimodal claims of transparency, transgression and originality, Vice.com builds a coherent narrative as a countercultural lifestyle media outlet. From a political economy perspective, however, the website's appearance points to more sinister implications. Vice Media has increasingly become controversial because of the discrepancy between its countercultural 'style' and the corporate 'substance' of its economics and ownership structure. In addition, Vice News in particular has been criticized for promoting problematic if not reactionary views on a variety of issues, while also being heralded as a more transparent and accessible news outlet.

On one hand, Disney and Murdoch's substantial financial investments in Vice Media, together with its most recent valuation at nearly US$6 billion, raise questions about the veracity of the media company's anti-establishment attitude. Along the same lines, Vice Media's approach to partnering with lifestyle brands like North Face or Lululemon to produce documentary content seems quite contradictory. However, this is also an approach that caters well to an audience demographic, that of Millennials, for whom lifestyle politics and consumption are clearly not at odds with each other. On the other hand, Vice journalists' immersive, first-person approach to reporting is often riddled with sensationalism and stereotypes – from a keen focus on accounts of cannibalism and alcoholism in African countries like Liberia and Uganda to attention-getting stunts like the carefully orchestrated and widely publicized meeting between former NBA star Dennis Rodman and North Korea's dictator Kim Jong-un (Stelter, 2013). Arguably, this **mixing of registers** (immersion and sensationalism) is strategically tied to a need to reach and please an audience identifying with the media outlet's anti-establishment attitude while also catering to the conservatism of Vice's corporate investors. Vice.com's sleek design paired with its grungy imagery is key to the communication of a contradictory (and therefore perhaps more transgressive and original) ethos.

In selling counterculture to large corporations like Disney and Murdoch's 21st Century Fox while also producing branded content aimed at Millennials, Vice Media is clearly at the forefront of lifestyle commodification. In recent years, however, Vice News has also been in the spotlight for its embedded journalism in the Ukraine and among ISIS fighters, which was deemed both groundbreaking and courageous by many news media professionals. The late *New York Times* columnist David Carr had famously criticized Vice founder Shane Smith for claiming that Vice News reporters' gonzo-style journalism surpassed that of mainstream news outlets in that these never told the 'whole story'. Four years later, Carr praised Vice News journalists for their willingness 'to do the important work of bearing witness, the kind that can get you killed if something goes wrong' (Carr, 2014b). In the end, Vice's historic beginnings as a punk magazine and its following rapid corporate ascent may both be at work in the production of some of the contradictory claims that animate its visual appearance.

CASE STUDY 2: FEMINIST STOCK PHOTOGRAPHY IN THE LEAN IN COLLECTION BY GETTY IMAGES

In 2014, Getty partnered with Facebook Chief Operating Officer (COO) Sheryl Sandberg's Lean In Foundation to curate a collection of images based on the key principles of her bestselling feminist manifesto *Lean In: Women, Work and the Will to Lead*, a book targeted at women aspiring to high-level, successful careers (Sandberg, 2013). Sandberg's book has been subjected to criticism because of its 'neoliberal' ethos, as this version of feminism promotes an entrepreneurial approach to developing 'creative individual solutions' (Rottenberg, 2014: 422) aimed at succeeding in the corporate world while also achieving a perfect balance between career and family – for example, by ensuring that male partners are equally and fully involved in household chores and child-rearing.

With the motto 'You can't be what you can't see', the stated aim of the stock images included in the collection is to 'shift perceptions, overturn clichés, and incorporate authentic images of women and men into media and advertising' (LeanIn.org/Getty). Following some of the key principles of Sandberg's bestselling book, the collection's specific aim is to represent women as both empowered and authentic. This double aim was made clear across official descriptions of the collection on the Lean In Foundation and Getty Images websites as well as press releases, webinars and other promotional materials about the collection (Grossman, 2015).

Research questions and approach

In this case study, we ask the following questions:

- How is female empowerment communicated in these images?
- How is female authenticity communicated?
- Overall, what kind of feminism is promoted through the Lean In Collection?

Because stock images are made to be used in the creation of other media texts, our analysis here is based on multiple methods and forms of evidence. Through a discourse and content analysis, we examine the visual **representation** of gender and feminism in the Lean In Collection. We then extend our findings by interrogating the ***production***, ***circulation*** and ***recontextualization*** (or uses) of these images. We draw from a set of in-depth interviews with Getty photographers, which allow us to integrate considerations regarding their creative process and the models' labour into a critique of the representational import of images from the collection. Through our collaboration with the Digital Methods Winter School in Amsterdam, we were also able to add quantitative information about the collection as a whole to our content analysis, while also tracing some of the ways in which these images are circulated and used in online media texts spread across social media, magazines and newspapers. Erik Borra, Donato Ricci and Federica Bardelli were largely responsible for devising our digital methods protocol for this case study (a step-by-step explanation can be found at: https://wiki.digitalmethods. net/Dmi/WinterSchool2016CriticalGenealogyGettyImagesLeanIn). Overall, this case study offers an example of how visual analysis can often benefit from an **integration of multiple approaches** – both textual and contextual, semiotic and sociological, qualitative and quantitative – when key research questions cover a range of issues and sites of investigation.

Findings: an analysis of representation, production, circulation and recontextualization

Discourse analysis: feminism as empowerment and authenticity

As a first step in our visual analysis, we looked at the Lean In Collection as a whole to identify the key visual themes that underlie the collection's claims of female empowerment and authenticity. In Lean In, **empowerment** is visually portrayed through four main themes: leadership, agency, togetherness and 'flipping' the script. From a visual standpoint, the portrayal of female leadership is largely achieved through the vectors (Kress and van Leeuwen, 2006) formed by a woman's hand gestures as she explains something or points to a particular detail on a board or monitor, together with the gaze lines of other subjects in the image, who usually all look in the direction of the main female subject or the detail that she is showing. In sum, there are many images of women explaining or showing things to others, especially men, and often to groups of people.

Across the collection, women's agency is emphasized through a visible display of actions, skills and muscles. Many images portray women holding tools, making things and using their bodies to perform challenging physical tasks. Interestingly, there is also a fairly large group of images that show women implicitly or explicitly cooperating and bonding. Most often, these images are tagged with the keyword 'togetherness'. Here, women are typically portrayed as equal members of a group (e.g. a football team, a professional team). In many cases, there are interactions like embracing, laughing or taking selfies together, but in other cases women are simply portrayed as performing the same actions (e.g. lifting weights, working on a computer, running) next to each other or in symmetrical arrangements that visually point to connotations of similarity and connection.

Another theme that is explicitly associated with the notion of empowerment across the collection and related promotional media is 'flipping the script'. In this case, women are portrayed in occupations stereotypically associated with men (e.g. firefighters or robot engineers), or as tomboyish or physically powerful. Just as openly, the collection portrays men in 'feminine' roles, like parenting, nurturing and caring.

As far as the collection's second key theme goes, a first approach to the visual communication of **authenticity** relies on the portrayal of 'ordinary' women, rather than typical professional models. These are often portraits, which may be shot in a real-life setting (e.g. a classroom, a storefront) or against the decontextualized background of a studio. Tied to the theme of authenticity is also an attention towards diversity. The collection includes a greater amount of portrayals of non-white, working-class, disabled and older women than your average stock archive. Another distinctive feature of these images is a tendency to place women in highly textured environments, like lush green fields or cluttered

FIGURE 10.4 Authenticity in the Lean In Collection as the visual communication of diversity, texture and reality. Source: Getty Images. Reproduced with permission.

workshops (Figure 10.4), which confer a tangible and grainy quality to these visual narratives (Aiello and Pauwels, 2014).

Another kind of texture that is emphasized in these images is the texture of the body (Aiello and Woodhouse, 2016). Unlike classic stock photos, these images often highlight the wrinkles of ageing models and include models with large tattoos. Finally, in these photos authenticity is sometimes tied to reality in its own right. The Lean In Collection also includes photos of real scientists, 'makers' and professionals portrayed as they do what they do in real life. We will return to this last point about 'real' stock photography models in a moment.

Content analysis: empowerment without diversity?

Starting from our broader discourse analysis, we examined the first page of the collection on the Getty database, which was set on the 'best match' search principle by default and consisted of 60 images. Our content analysis helped us link some of the broader visual themes identified earlier with a more in-depth discussion of the representational characteristics of these images. In examining the collection's first 60 images, we focused on categories of manifest content such as **activity**, **setting**, **pose** and **gaze**. We also combined this focused, qualitative content analysis with digital methods aimed at gathering information about the collection as a whole. This helped us gain an understanding of generalized patterns of representation regarding age, race and physical appearance, together with an understanding of the collection's key conceptual underpinnings.

Of the 65 women who were portrayed across the collection's first 60 images, only five were over 55 years old and less than a third were visibly non-white. Researchers from the Digital Methods Winter School were able to 'scrape' the most frequently used **keywords** in the collection used to describe the physical characteristics of portrayed subjects. What stood out immediately was an overrepresentation of white people. For every black woman, for example, five Caucasian women were represented in the collection. Overall, the typical person depicted in a Lean In photograph was a 25-to-29-year-old Caucasian woman with long brown hair.

Over a third of the collection's first 60 images showed women at work and/or using technology (e.g. smartphones and digital tablets) in an office space, a home office, a classroom, or an extension of the office such as a café. Other images showed women in a mechanic or artisan workshop, a science or engineering lab, at the gym or in a swimming pool, in a café or more generally outside – on the street, in a playground or on a rooftop garden. The few instances in which women were portrayed in a domestic setting correspond to parenting as the key activity being represented. Only 5 per cent of images in our qualitative sample portrayed a woman parenting a child in the absence of a partner, while 10 per cent of these images portrayed men caring for children in the absence of a woman.

The 'waist up' and 'front view' keywords are also relevant here, as most of the collection's images were clearly shot from a close range and showed the face of the person photographed, indicating that there tends to be an emphasis on the person as an individual with particular physical attributes rather than any group or environment that she may belong to. Not surprisingly, then, 25 per cent of the collection's first 60 images were portraits of individual women or girls – that is, images that were clearly shot without an intent of representing an action or narrative, but rather emphasizing an individual's characteristics or 'typical attributes' (Kress and van Leeuwen, 2006). Almost half of these portraits were shot in an office environment, whereas only three of them were classic studio shots in line with the 'traditional' stock photography aesthetic.

On a conceptual level, however, an emphasis on being together was also evident. Keywords like 'togetherness', 'connection', 'bonding' and 'cooperation' were most frequent in our digital content analysis. Along the same lines, 25 per cent of the images in our qualitative sample portrayed pairs or a small group of women interacting with each other (e.g. laughing or taking selfies together) or women collaborating with men. This is an interesting finding because one of the main criticisms of Sandberg's version of feminism as a lifestyle is its focus on the individual, and on individualized goals and solutions.

Happiness and business-related concepts were clearly represented, something that is in alignment with more traditional approaches to stock photography where imagery must be upbeat and is often linked to the world of creative and corporate work (Machin, 2004). Photos tagged as 'daytime' and 'indoors' were prevalent, a finding that is also tied to the prevalence of work-related photos shot in offices during the day.

In sum, 50 per cent of the images in our qualitative sample showed women in a work setting. Equal amounts of images show women as parts of groups and women as individuals. In addition, the titles given to the photos were analysed through digital methods. Among the top ten most frequently used words within these titles there were terms related to women ('woman', 'female', 'girl', 'businesswoman') and work ('office', 'businesswoman', 'working'). Also, the words 'young', 'smiling' and 'portrait' stood out. These results indicate that a typical image found within the collection portrays a young, smiling woman at work (Figure 10.5).

FIGURE 10.5   A 'typical' Lean In woman: young, white, smiling and at work in an office environment. Source: Getty Images. Reproduced with permission.

As a whole, it seems clear that the collection's version of empowerment is heavily premised upon ideals of corporate success, individual distinctiveness and collaborative productivity. Work–life balance also plays a role here, as about a third of the images depicting women focus on leisure, parenting or non-professional achievements. Ultimately, however, Lean In's version of feminist empowerment is also largely rooted in a general lack of diversity and an overall orientation towards corporate business.

Looking beyond the frame: the Lean In Collection in context

We examined findings from our discourse and content analysis against evidence regarding the production, circulation and recontextualization of Lean In images. These are **contextual factors** that actively shape the visual substance and representational politics of images from the Lean In Collection.

Many of the 40 photographers we interviewed in relation to their stock imagery work spoke about the importance of trends tied to popular digital platforms like

Instagram (Aiello, 2016). They pointed to the perceived need to craft images that look as if they have been shot with an iPhone and, as one of the photographers said, also feel 'less staged and produced'. The photographers also talked about how stock photography has become 'a lot more documentary, a lot more realistic' with the rise in popularity of amateur styles of photography that privilege ordinary people and real-life moments.

We interviewed one of the photographers who used 'real' women in their real-life occupations as subjects for their stock photos, focusing specifically on portraying scientists at work in their labs. We asked her about how she recruited actual scientists as models, and she said that they responded enthusiastically because they wanted 'to inspire the next generation'. The models were not given money, but only profile shots, and they helped the photographer keyword the images for accuracy. However, given the freelance nature of the photographer's work, there was no guarantee that the photos she took of them would be included in the Lean In Collection. Finally, there was a lack of control and overall uncertainty regarding the final uses of these images, which could be purchased by virtually anyone and for any message, including big pharmaceutical companies, for example.

As a whole, our interviews show that the Lean In Collection linked to attempts to profit from the adoption and promotion of commercially popular visual styles associated with social media. Furthermore, both the photographers and the models involved in the creation of these images often have little control over the conditions and the outcomes of their labour.

By using the Digital Methods Initiative's Google Reverse Image Scraper (which is freely available at: www.digitalmethods.net/Dmi/ToolGoogleReverse Images), we were also able to understand where most of these images 'live' and how some of them are used in particular media texts. Overall, there was a disconnect here between some of these images' subject matter and both their distribution and uses which, in the end, were not regulated by the stated aims of the collection as such. In terms of circulation, Lean In images were mostly found in social media, with Pinterest and Twitter right at the top (Figure 10.6).

In most cases, these images are used for lifestyle media content like fashion, food, sports and health. A significant finding here is that Lean In images are not used frequently in tech journalism, even when they portray women using or building technological devices. Another meaningful finding is that photos of non-white women tend to be used on websites that discuss issues of race and ethnicity and are focused on black communities, immigrants or Muslims. Likewise, we found that images of female scientists are mostly used in media texts about women in science, rather than in articles about science more generally. And often, these media texts emphasize the abnormal or obstacle-ridden nature of a career in science for a woman. For example, *Cosmopolitan* used a Lean In image of a female scientist to illustrate a story about Eileen Pollack, a physicist who left

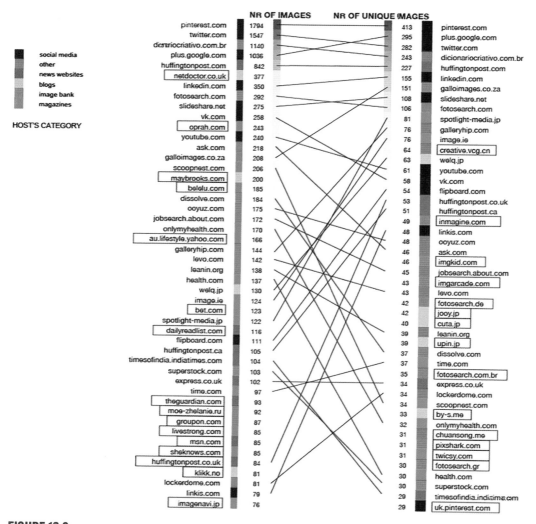

FIGURE 10.6 A visualization of the number of unique images from the Lean In Collection per host. Source: 2016 Digital Methods Winter School.

science in the 1970s because it was too much of a 'boys' club', only to conclude that to this day not much has changed (Figure 10.7).

Case study conclusion: is the Lean In Collection feminist?

Our discourse and content analysis have highlighted that the collection's approach to both empowerment and authenticity is limited and in fact orientated

FIGURE 10.7 A Lean In image of a scientist in a *Cosmopolitan* article about the struggles of women in science.

towards lifestyle ideologies that are not far from the **neoliberal** principles that Rottenberg (2014) and other critics identified as central to Sandberg's feminist manifesto. As a whole, the Lean In Collection promotes a feminist lifestyle ideology based on a version of womanhood that is largely confined to younger white demographics and that highlights individual achievements, corporate success, nuclear family models and middle-class forms of leisure. In the collection, there is very little emphasis on feminists' collective and intersectional approaches to tackling everyday politics.

However, the collection is also fairly non-stereotypical for stock photography as a visual genre. Women are represented as more agentful and as more diverse than in other areas of commercial imagery, despite a relative lack of diversity in terms of age, race and occupation and, by association, also class. Lean In does make an effort to portray women as leaders and men as supporters and nurturers, while also promoting both the individual qualities and collaborative skills of women and girls. The collection also endeavours to communicate a more complex picture of women's identities beyond their looks and as being clearly tied to everyday activities and environments. Along the same lines, it also includes imagery of women portrayed in their real workplaces. Finally, these images feature a varied range of body types and physical attributes like wrinkles and natural hair.

Overall, it could be argued that the problem here is not so much with the visual content of these images per se. Of course, this must be interrogated and the collection's representational pitfalls must be highlighted, as we did in this case study. Clearly, the collection's visual politics have a lot to do with the commercial interests, labour issues and aesthetic shifts of stock photography as a genre and an industry. When we look beyond the images themselves to explore their production and uses, for example, little seems to have changed.

Photographers are powerless in relation to the final uptake of these images, and the women portrayed in Lean In images may very well be performing forms of labour that are not properly remunerated. At the same time, Lean In images are still used to illustrate 'lightweight' lifestyle subject matter and, on the whole, are relegated to discursive ghettos. Ultimately, our analysis of the Lean In Collection suggests that, when it comes to feminist lifestyle politics, there is still a lot of work to be done in media culture at large.

CHAPTER CONCLUSION: LIFESTYLE IMAGERY BETWEEN TEXT AND CONTEXT

In this chapter, we have examined some of the ways in which lifestyles can be 'sold' visually to audiences across a range of media outlets and genres. In particular, our case studies have focused on the **commodification of lifestyle politics** in commercial media. Across the two case studies, we have noted that a critical evaluation of given versions and visions of lifestyle ideologies like counterculture and feminism is not to be reduced to the media texts' 'appearance' in itself, but ought to be grounded in a nuanced approach to linking visual texts with the broader contexts in which they are produced, distributed and used. As a whole, this chapter's key takeaways are threefold:

- By analysing the ways in which lifestyle ideologies are communicated through multiple types of visual resources, and in particular through stock images and web design as key to lifestyle communication, we are able to understand how contemporary media are increasingly in the business of promoting a range of ways of being and living which may reach well beyond mere consumerism.
- We can start to understand how visual analyses addressing both the textual and contextual dimensions of visual communication may offer more nuanced findings and therefore also allow us to reach conclusions that we may not expect from examining the text or the context alone.
- A variety of methods enable us to gather evidence on the context(s) in which images are produced, distributed and consumed. These include political economy, interviews with producers (or, conversely, audiences) and digital methods that can help us collect evidence on how imagery circulates and is used across different media platforms.

> **POTENTIAL FURTHER RESEARCH**
>
> This chapter has focused on imagery that promotes and exploits lifestyle politics based on feminist and countercultural values as an integral part of corporate media. The analyses we proposed here could be extended to a variety of media rooted in 'alternative lifestyles', such as, for example, yoga magazines or vegan lifestyle blogs. A key question here could be: 'How do these magazines or blogs use a combination of visual styles to promote particular ideological values and consumer practices?'
>
> However, another way to explore the significance of lifestyle imagery for audiences would be to conduct research with those who consume these images – for example, readers of men's lifestyle magazines or celebrity lifestyle websites like Oprah.com or Goop. In addition to analysing the visual style and content of a magazine or website, guiding questions here could be: 'How do audiences describe and interpret the different kinds of imagery that are used in these magazines or websites?' 'Is the style or aesthetic of a magazine or website central to the ways in which members of an audience make decisions about how and whether to follow particular lifestyle guidelines?'

FURTHER READING

For social semiotic approaches to analysing lifestyle imagery, see Machin and van Leeuwen (2007), Adami (2015), Aiello and Woodhouse (2016). Lewis (2008) and Bell and Hollows (2005) are useful sources on lifestyles in the media and popular culture.

11 BRANDS AS VISUAL EXPERIENCES

Branding is key to how products and services are communicated through the media. Brands are also central to how we experience our relationship to particular goods, institutions and companies in everyday life. For this reason, in this chapter we consider brands not simply as names or identities for consumer goods, but rather as meaningful relationships between commodity providers and consumers or users. We examine the visual dimensions of branding, particularly logo design, to gain insight into how imagery and other visual cues are used to sell commodities, and how media culture contributes to the development of particular forms of attachment between people and brands. In sum, this chapter will:

- review definitions of branding and approaches to promotional culture that offer critical and empirical insights about the experiential dimensions, social significance and cultural meanings of brands;
- explore the role of visual communication in the creation of brand identities and in framing the values and qualities that set them and their potential users apart;
- show how rebranding works from a visual and multimodal standpoint, and how, over time, changes in a popular sharing economy brand like Airbnb correspond to major shifts in how its users are imagined and positioned (Case study 1);
- examine some of the ways in which users' perspectives on the aesthetic appeal of logos may contribute to the visual analysis of a selection of global digital media technology and communication brands, including, for example, Instagram, Apple and Facebook (Case study 2).

The chapter's case studies focus on digital commodities that are just as immaterial as they are staples in the everyday lives of many. This emphasis on **intangible goods** and online services enables us to highlight further the importance of the symbolic and aesthetic aspects of branding in framing something as a saleable commodity with distinctive and desirable qualities. Before we address the paramount importance of visual communication in this process, in the next section we outline key theoretical concepts and research findings related to branding more generally.

BRANDING AND BRANDS BEYOND ADVERTISING

The first known product to be branded was a bar of soap produced by the American company Procter & Gamble, also known as P&G. At the end of the 1870s, Harley Procter decided to replace the soap's generic denomination as 'P&G White Soap' with the name 'Ivory Soap', which apparently came to him after reading the Bible (Sivulka, 2012). This decision to give the P&G soap a unique 'brand' name was followed by distinctive choices in matters of packaging and promotion. Ivory Soap was wrapped in white paper featuring an elegant blue design instead of the plain brown paper that was typically used at the time. The soap was also promoted through the famous slogan '99 and 44/100% pure', and was advertised nationally through attractive illustrations.

The history of branding is entwined with that of advertising, and yet these are by no means the same thing. While one of the main aims of advertising is to create awareness, branding aims to foster long-term engagement and **loyalty** among customers. We examined advertising in greater detail in Chapter 10, so here suffice it to say that branding is aimed at making commodities both distinctive and desirable, and that branded products and services are then typically advertised through traditional and social media alike, but also that they may not be advertised at all. For many years, a globally famous brand name like Starbucks chose to forgo advertising campaigns, focusing instead on building the coffee company's brand image through word of mouth, through events and, even more so, through their stores (Schultz and Jones Yang, 1997). At the same time, an even more ubiquitous brand like Coca-Cola systematically built its brand image through advertising, both in the media and in public space. In addition to appearing in television and radio commercials as well as on branded objects, billboards and even as hand-painted signage on building façades the world over, the Coca-Cola brand has profited from sponsoring heavily mediatized global events like the World Cup and the Olympics (Lash and Lury, 2007). In sum, most often branding and advertising go hand in hand, but branding operates at the level of a commodity's core values and qualities.

Since its inception, branding has therefore expanded to all aspects of a commodity's presence in the world. It is in this sense that brands are **experiences**, rather than simple labels. This said, the basic act of naming saleable goods is still quite powerful in its own right, as it 'humanizes' them by adding 'the same kinds of meanings that are reserved for people' (Danesi, 2006: 14). What this entails is that, at their core, brands confer a personality and indeed an identity to inanimate objects and immaterial entities. In doing so, they 'link acts of consumption to social relationships' (Davis, 2013: 4).

Building a successful 'brand image' has become an imperative across different areas of social life and 'the economic success of products, companies, nations and even individuals' (Fairclough, 2002: 163) now depends largely on branding. Major branding agencies are in the business of crafting effective brands for alcoholic beverages, cars, hotel and restaurant chains, airlines and sports events, just as much as they are asked to create brand identities for non-profit organizations, cities and countries (Aronczyk, 2013). By now, it has also become standard practice for politicians, celebrities and social media influencers to work on the creation of their personal brands with public relations consultants and marketing agencies.

It is in this sense that branding is not confined to proper commodities as such. Rather, branding is what Moor (2007) defines as a 'technique of governance', because it involves concerted efforts to influence the perceptions, behaviours and overall **attitudes of publics**, whether these are made of customers or citizens. It is therefore not surprising that a global leader in brand consulting and design such as Landor Associates has had clients that include Kraft, Nike, Barclays, HSBC, Samsung, Microsoft, BMW and Cathay Pacific, but also the City of Hong Kong, the City of Melbourne and the Sultanate of Oman. Overall, then, brands are what Lury defines as 'an *interface* of communication between producers and consumer' (2004: 48). In establishing a relationship between producers and consumers, or users, brands also become a form of currency (Arvidsson, 2006). As we will see in the rest of this chapter, much of this ability that brands have to establish meaningful relationships with customers and acquire economic value in a given marketplace is achieved through visual means, regardless of how similar their products or services may be to those of other brands.

THE VISUAL DIMENSIONS OF BRANDING

Branding draws from widely available cultural or 'semiotic' material to transform products, services, organizations, countries, cities and even individuals 'into mental phenomena' (Danesi, 2006: 8). Hence, the notion of brand image refers broadly to the various **cultural associations** that a brand evokes in consumers, or the ways in which various meaning-making resources are organized in a cohesive whole that conveys multiple, related connotations. The resources that are used to develop a brand's image are by no means only visual, and they usually encompass a wide variety of communication modes and sensorial stimuli. This said, the visual dimensions of branding are particularly significant, both because they are multifaceted and because they can be more easily reproduced and circulated across media.

Because a single brand often acts as a set of relations between different products and services, and can therefore also generate multiple goods (Lury, 2004; Lash and Lury, 2007), visual cues become especially important in developing and maintaining a cohesive identity for the brand. As a somewhat extreme example of this property, the Virgin brand encompasses a wide variety of businesses that include banking, travel and hospitality, media and entertainment, healthcare and fitness, and digital technologies and communications. Funded by the eccentric British business magnate Sir Richard Branson, Virgin prides itself on having a 'backbone' that ties all of these services together under the umbrella of values like 'providing heartfelt service, being delightfully surprising, red hot, and straight up while maintaining an insatiable curiosity and creating smart disruption' (Virgin, n.d.). The visual cues that are key to the Virgin logo and overall aesthetic, like the colour red and the handwritten or spray-painted signature, contribute to building and reinforcing Virgin's image as an off-the-wall, maverick brand.

As a whole, successful branding requires consistency, or what is usually defined as brand continuity, particularly when the commodities that are being sold under the same umbrella are diverse, as in the case of Virgin. **Brand continuity** entails a coordinated approach to the 'message', where all elements of the brand – logos, typefaces, packaging, writing styles, and even smells and sounds – ought to feel as if they are cut from the same cloth and therefore also contribute to a seamless look and feel. Brand continuity leads not only to greater brand recognition, but also to a greater degree of trust among potential and actual customers.

Nike is a perfect example of how a successful brand is built on carefully orchestrated continuity. Across channels and media, the Nike brand maintains a consistent message that is firmly centred in the globally famous Swoosh logo and 'Just do it' slogan. Nike advertising campaigns have featured star athletes like Michael Jordan and Tiger Woods, as well as ordinary people and street artists, while consistently offering narratives centred on individuals' heroic pursuits and personal transcendence (Goldman and Papson, 1998). Along the same lines, Niketown stores combine spatial and interactive elements borrowed from theme parks and museums, such as playgrounds and displays of celebrity athletes' memorabilia, which position visitors as active participants and heroic individuals in their own right (von Borries, 2004; Peñaloza, 1998).

Starbucks is another good example of a strong global brand that spreads its core message across different means of communication to achieve a unified brand identity. The logo, packaging and store designs are all coordinated to convey core values like authenticity and community – for example, through key colours like green and brown, soft and circular shapes, and iconographic motifs like the siren and signature swirls or 'handwritten' text (Aiello and Dickinson, 2014).

Across these examples, the visual dimensions of branding play a crucial role in building a brand's image, and in keeping it both consistent and novel. Brands, then, can be defined as visual experiences to the extent that their visual elements are integral to the creation and maintenance of a unique relationship between the brand and its users through and across multiple platforms, including television, print media and social media.

THE POWER OF LOGOS

When it comes to evaluating the attractiveness of any given brand, it is logos that take centre stage, to the extent that often the notion of 'brand' is confused with that of 'logo' itself. Indeed, logos are the 'face' of branding; while the brand in itself is 'intangible or incorporeal', a logo is what 'makes the brand visible' (Lury, 2004: 74). Generally speaking, logos may be seen as fulfilling two main functions. On one hand, a logo is an **index** of the brand's producer, as it is both a unique and reproducible trace that points to the authenticity and quality of the branded commodity. On the other hand, logos are also **iconic**, as they communicate qualities that are supposed to resemble the brand itself. It is not unusual for a logo to be planned and described through a series of adjectives (e.g. 'friendly', 'dynamic', 'refreshing') that may be attributed to the brand as a whole.

Corporations, organizations, and even countries and cities rely on logos to position themselves and succeed in a variety of marketplaces. For example, after its handover to China in 1997, Hong Kong created a brand to promote its 'world-class' reputation, mainly with the aim to keep attracting international business and tourism. This intention was made visible through the adoption of the slogan 'Asia's World City' and of a logo designed by Landor that featured a stylized, aerodynamic dragon, which was later updated to include three colourful ribbons pointing to Hong Kong's dynamism and diversity.

Not surprisingly, then, both branding and rebranding are linked to systematic choices in the design or redesign of logos. As far as **rebranding** goes, it is common for major corporations to 'rethink' their brand by tweaking or redesigning their logo first. Since its first appearance in 1971, the Starbucks logo has been through at least four major redesigns, with the logo being increasingly stylized to tame the sexualized traits of their signature siren, while also making her look more stereotypically feminine, both by 'covering' her breasts with her hair and giving her more streamlined facial features. In its most recent redesign, the logo has lost its outer circle with the wordmark 'Starbucks Coffee', thus communicating Starbucks as a brand that is no longer simply mainstream and family-friendly, but also instantly recognizable and about much more than just coffee (Aiello, 2018).

From a visual communication standpoint, then, logos are particularly powerful for three main reasons. First, where brands are 'signs that stand for ideas with a strong emotional appeal' (Danesi, 2006: 17), logos are pictorial signs that stand for brands and, for this reason, their visual design features are often associated with symbolic if not **mythical traits** that tap into our deepest values, desires and aspirations. Among other things, for example, Apple's success is tied to its logo's ability to evoke the biblical story of Adam and Eve, where Eve, the mother of all humans, eats the fruit that contains forbidden knowledge (Danesi, 2006). The Apple logo also evokes Isaac Newton's stroke of genius, as the story goes that he came up with his theory of gravity when an apple fell on his head while he was sitting under a tree in his garden (Floch, [1995] 2000).

Second, as Christian Mosbæk Johannessen (2017: 2) states, logos are 'structurally minimalist' and their **simplicity** is more 'apt for conveying what logo designers refer to as *look and feel*'. Once again, the Apple logo is a good example of how simple shapes and minimalist details can convey a nuanced message. Finally, logos are at the centre of a 'a culture of circulation' (Lash and Lury, 2007: 132), as they are the main form of **symbolic currency** that is exchanged and disseminated in promotional communication, most often through global flows of branded objects, but also via digital media. In other words, a brand's success is most often tied to the ubiquity of its logo.

In this chapter's first case study, we examine the branding and rebranding of Airbnb to understand how key visual and linguistic resources have changed over time to promote this prominent sharing economy brand, first as a functional marketplace for more affordable alternatives to hotels, and later as a lifestyle community. The second case study is a reception-led semiotic analysis of the logos of a sample of popular digital media and technology brands. Here, we examined the views of a group of participants on how particular visual cues (e.g. colour and shape) contribute to their judgements and feelings regarding logo design in order to understand how they evaluate and affiliate with different brand identities.

CASE STUDY 1: THE POLITICS OF AUTHENTICITY IN THE REBRANDING OF AIRBNB

Airbnb was founded in 2008 by three San Francisco entrepreneurs, Joe Gebbia, Nathan Blecharczyk and Brian Chesky. By 2011, after a series of major investments by venture capitalists, Airbnb was reportedly valued at US$1.3 billion. By 2012, it had hit over 10 million bookings (Salter, 2012). As it kept expanding to 65,000 cities, Airbnb also became controversial for the negative impact that its short-term rentals have on housing provision, living costs, and ultimately also displacement and gentrification.

On 16 July 2014, Airbnb launched its new brand identity with great fanfare. With a new logo and an overhaul of its app and website, the popular online marketplace for short-term accommodation aimed to promote a new focus on 'belonging' as part of a 'community of individuals'. When the new brand identity was launched, *Wired* highlighted how Airbnb's new 'look' intentionally emphasized 'fewer of the homes you might stay in, and more of the lifestyle you might have' (Kuang, 2014). A host of other commentators in design and advertising trade magazines hailed Airbnb's rebranding as a move towards linking the brand with people, experiences and community, rather than technology and economics alone. In the wake of legal disputes, government crackdowns and a number of protests in cities like Barcelona and San Francisco, the rebranding of Airbnb was certainly a strategic move to promote an image of the company as caring and community-oriented.

Research questions and approach

In this case study, we ask two main research questions:

- How has the Airbnb brand changed over the years, and how have particular visual resources been used to communicate different meanings in relation to the brand at different times?
- What does the visual rebranding of Airbnb tell us about the changing cultural and political implications of this global brand?

Here we adopt a **social semiotic approach** to data collection and analysis, which highlights the relationship between changes in the semiotic resources found in particular texts (e.g. homepage screenshots) and the broader economic, cultural and professional factors that shape them. Using the Internet Archive's 'Wayback Machine' (a database that systematically captures cached pages of websites and stores them in a fully searchable digital archive), we collected a sample of 36 homepage snapshots from the Airbnb website between 2009 and 2018. However, these archived web pages, particularly older ones, often miss stylesheet information (CSS) and/or images. For this reason, we also collected a number of web pages and media articles about the Airbnb brand and its rebranding. We entered the phrases 'Airbnb brand' and 'Airbnb rebranding' into a widely used search engine to find additional evidence regarding changes in the appearance of the Airbnb website over time. In addition, we mined a selection of media articles for commentary by professional critics (e.g. journalists, designers) and by those responsible for the look and feel of the brand itself, including Airbnb's leadership, marketing managers and designers.

Findings: from functional and community brand to intimate and lifestyle brand

We now turn to our social semiotic analysis of the four main stages in the life and evolution of the Airbnb brand. Here, we discuss how the **representational**, **interpersonal** and **compositional** meaning potentials associated with key semiotic resources and their combinations contribute to defining Airbnb's brand identity across these different stages (see Chapter 2 for an explanation of social semiotic analysis). In offering a descriptive and interpretive analysis of these resources and meaning potentials, we will explain that, over time, Airbnb has been communicated as a functional, community, intimate, and lifestyle brand. We will then use this analysis for a critical discussion of the implications of Airbnb's move towards the promotion of authenticity as integral to its brand identity.

2008–10: Airbnb as a functional brand

Airbnb was originally named 'Airbed and Breakfast' with the tagline 'Forget Hotels'. By the end of 2009, the new name 'Airbnb' and a new logo had been introduced, together with the new tagline 'Travel Like a Human'. In this early phase of Airbnb's life as a brand, the representational meanings of its homepage were related to the website's functional offerings. The overarching 'story' communicated here was one of utility and convenience. The search function's design was not unique to Airbnb, and the search function was directly associated with linguistic messages such as 'Find a Place to Stay' and 'Where are you Going?'. Website imagery was shot in a naturalistic way, seemingly to describe available

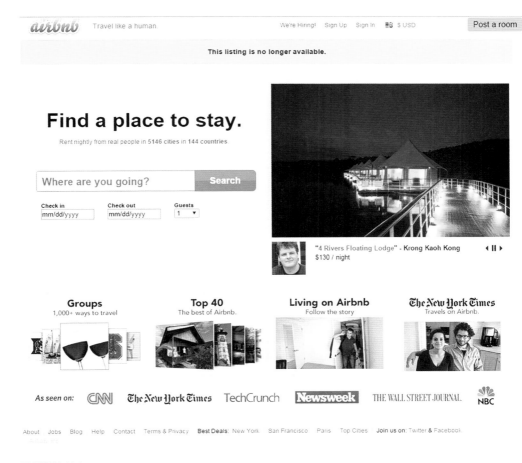

FIGURE 11.1 Airbnb website homepage in July 2010. Source: www.cbinsights.com/research/unicorn-websites-kinda-ugly-sometimes/

FIGURE 11.2   Airbnb's first logo.

rentals in a straightforward manner. The website addressed users both directly and impersonally at the same time, in that its perfunctory linguistic messages were most often matched by medium to long shots of available rentals' exteriors. From a compositional standpoint, the homepage was divided into different areas of content, where the key search function was clearly communicated as prominent, as it was placed on the top left area of the homepage, and the linguistic text that was made to look most salient was 'Find a Place to Stay'. The website's main interactive functions were largely independent of images of rentals and hosts, which were presented in separate 'boxes' of content framed by grey borders (Figure 11.1).

Airbnb's founders talked about the company's early logo design and overall branding as something that had been created in just a few hours for temporary use (Carr, 2014a). Both the Airbnb website and logo were designed to communicate a straightforward brand identity based on a few fundamental characteristics and functional goals like being able to look up different options for accommodation. Overall, Airbnb's early logo design was closely associated with the brand's name as such (Figure 11.2) and was in line with the design choices of now major tech brands like Facebook, Twitter and LinkedIn, relying on lettering that pointed to the brand's name and different hues of blue, a colour which in Western visual culture has become associated with wisdom, reason and trust, hence becoming a favourite colour in corporate culture (Pastoureau, 2001).

2011–13: Airbnb as a community brand

In 2011, Airbnb started encouraging hosts to post high-quality images of their rentals and even launched a free photography programme that allowed them to schedule a photo shoot of their homes with professional photographers. In a blog post, Airbnb stated that the photography programme was aimed at its 'community' and claimed that 'being able to offer photos and making a personal connection with the hosts was transformational for Airbnb' (Airbnb, 2011). A greater amount of pictures of interiors started appearing on the website. These photographs were not only naturalistic portrayals of hosts' homes, but were also typically shot in such a way – at a straight angle and at eye level – that potential guests could imagine being in the living rooms, kitchens or bedrooms that they portrayed. In other words, both the representational and interpersonal meanings of these images worked together to make guests and hosts feel more connected with each other and with Airbnb as a 'community'.

In the summer of 2012, Airbnb redesigned its homepage to feature large images of different rentals as backgrounds for its search function, therefore inviting potential guests to visually engage with some of their listings while searching for accommodation. From a compositional standpoint, this design choice brought website users visually closer to Airbnb's core offer (Figure 11.3). Along the same lines, Airbnb later introduced its 'Neighborhood Guides', which was presented on the homepage as a grid of thumbnail images portraying 'street scenes' from a selection of cities around the world. This feature aimed to create closer forms of engagement among Airbnb's community of users, as it allowed guests to research curated information about different neighbourhoods in any given city based on their tastes, interests and lifestyles (Lawler, 2012).

FIGURE 11.3   Airbnb website homepage on 1 August 2013. Source: Internet Archive Wayback Machine.

2014–15: Airbnb as an intimate brand

In 2014, Airbnb launched the new brand concept 'Belong Anywhere'. Airbnb's Chief Marketing Officer, Jonathan Mildenhall, stated that this rebranding aimed to 'create emotional and compelling stories that make people feel like they're having a human experience as opposed to being sold something by a brand' (Jack, 2015). This was a move to grow Airbnb's 'brand intimacy quotient', or Airbnb's ability to engage and bond with its users (Natarelli and Plapler, 2017). The new homepage featured the slogan 'Welcome Home' against the backdrop of photographic tableaus that came to life to portray a series of scenes which represent Airbnb guests exploring new cities, interacting with hosts, or enjoying some quiet time 'at home'. This living imagery kept renewing itself across the width of the browser while users could type in the destination and dates of their next trip (Figure 11.4). Across the homepage, there were images that focused on hosts and their homes, together with places and activities rather than the rentals as such (Kuang, 2014).

By mobilizing archetypes like 'home' and 'belonging' through visual imagery, the new Airbnb homepage promoted meaning potentials like warmth, closeness and connection. The immersive quality of both moving and still images augmented these representational meanings, highlighting the importance of personal engagement with the brand. This imagery worked to shorten the 'social distance' between Airbnb users and the website's offerings through medium shots and close-ups of both people and places. Likewise, the new tagline 'Belong Anywhere' and the new slogan 'Welcome Home' addressed Airbnb users directly, while also interpellating them to 'feel' an emotional attachment to Airbnb's values – for example, through direct appeals that addressed users as part of a community with shared beliefs and sentiments. On a compositional level, the website's key search function and the logo were juxtaposed to the living imagery, creating a close relationship between the act of searching for accommodation and the 'Airbnb lifestyle'.

FIGURE 11.4 Airbnb website homepage on 1 May 2015. Source: Internet Archive Wayback Machine.

The new logo (Figure 11.5) was named Bélo to evoke the concept of belonging while also sounding international (Jack, 2015). The logo was also designed to be customizable, so that anyone could reproduce it in their own way (Carr, 2014a). Its key hue, magenta, evoked the energy of a warm colour, while also being associated with meaning potentials such as modernity and technology like other flat colours (Kress and van Leeuwen, 2006). Together with the curvaceous appearance of the logo design, this colour choice was associated with meaning potentials like femininity and nurturing. Not surprisingly, Twitter critics and Tumblr parodies of the new logo emphasized its resemblance to a vagina (Wainwright, 2014).

FIGURE 11.5   Airbnb's 2014 logo.

As a whole, in this phase the Airbnb brand harnessed its established success both as a functional and a community brand to move into realms like intimacy and, as we will see in a moment, also lifestyle.

2016–present: Airbnb as a lifestyle brand

In 2016, Airbnb's main slogan changed from 'Welcome Home' to 'Live There'. Airbnb's largest marketing campaign to date promoted it as a brand that enabled travellers to 'live' in a variety of places across the world, just like locals do. To this end, Airbnb also harnessed the 'authentic' experiences of local hosts to create guidebooks that offered 'a taste of what day-to-day life is like for people who actually live in the city' (Richards, 2016).

By the end of 2016, the website had been redesigned, featuring a pared-down aesthetic that privileged typography and minimalist photography over immersive imagery. Phrases like 'Live there. Book unique homes and experience a city like a local' (1 November 2016), 'Stay. Play. Discover. Find unique homes, experiences, and local guides for your trip' (1 February 2017) or 'Airbnb Book unique homes and experience a city like a local' (1 May 2017 and 1 November 2017) dominated the top of the homepage. The main font, LL Circular, was sans serif and rounded, and visually echoed the logo's clean yet organic design. The homepage was now set apart by a white background dotted with the iconic magenta colour. Scrolling down the homepage, one could click on neatly composed images of Airbnb rentals.

From a representational or ideational standpoint (see Chapter 2 for a definition), this choice to focus on the brand's distinctive colour and shape established Airbnb's brand identity as key to user experience in its own right, thus associating the brand with users' attitudes and ways of being. From a compositional standpoint, the salience of both colour and shape points to the promotion of the very essence of the Airbnb brand as most significant in its communication. As of early 2018, this approach to designing the Airbnb website was still in place, but the homepage had been populated with more information and imagery pointing to the various 'homes' and 'experiences' that a wide base of users could access based on their tastes and needs – for example, according to whether they were travelling with family, for work or for particular forms of leisure like surfing, cooking or glamping (Figure 11.6). Overall, the brand was now centred on a lifestyle-oriented approach to using Airbnb.

FIGURE 11.6 Airbnb website homepage on 3 March 2018. Source: Airbnb.com

Case study conclusion: authenticity and the politics of the Airbnb brand

The notion of authenticity has long been central to branding, as effective brands ought to be able to communicate and evidence their originality, the trustworthiness of their values and their validity, or how 'true' the stories they tell about themselves are (see van Leeuwen, 2001). However, by now, brands are also part and parcel of everyday politics, identities and lifestyles, and authenticity has become central to how brands engage in the non-corporate, non-consumerist areas of our lives. Because what is understood as authentic is typically also 'perceived as *not* commercial' (Banet-Weiser, 2012: 10), authenticity in branding is also linked to attempts, particularly by global corporations, to establish genuine forms of engagement and connection with customers. We are therefore now often interpellated by brands as caring individuals, community members and even responsible citizens.

Over just a decade, key shifts in Airbnb's homepage reflect the increasing importance of claims to authenticity in the promotion of Airbnb as a brand. With regard to representational meanings, we see the growing inclusion of portrayals of home interiors, individual hosts and guests, and everyday places and activities. Interpersonal meanings are progressively centred on shortening the social distance between the viewer and the people and places represented in key imagery, thus conveying a sense of closeness that is augmented by the introduction of immersive visual styles, and both warm colours and organic shapes. Finally, compositional meanings progressively point to the centrality of Airbnb's core brand identity in the act of looking for accommodation and experiencing a place. The rebranding of 2014, in particular, brought authenticity to the fore as a way to promote stronger emotional attachment between Airbnb users and the brand. While Airbnb had already focused on community in building its brand, this major act of rebranding entailed semiotic resources that promoted a greater degree of intimacy between Airbnb and its community of both guests and hosts, who were often portrayed as part of a web of warm, close relationships.

As a tech giant now worth over US$30 billion, Airbnb can be hardly equated with informal, small-scale and ultimately also non-commercial approaches to the business of hospitality. Its rapid growth has gone hand in hand with the growing presence of serviced apartment providers, owners of multiple properties and landlords whose long-term rentals have been turned into full-time Airbnb accommodations, thus displacing tenants and altering the nature and costs of housing provisions in many cities across the world. Major protests in cities like San Francisco and Barcelona have foregrounded Airbnb's negative impact on local communities, due to the corporation's role in the dispossessions, evictions, and soaring housing costs caused by tourism gentrification (Iyengar, 2015; Burgen, 2017). Over the years, this increasing corporatization of Airbnb has, however, corresponded to a greater emphasis on notions like 'community', 'locality' and 'experiences', which are all associated with authenticity. The more impersonal its business has become, the greater an effort Airbnb has made to establish itself as a lifestyle brand by harnessing both the image and labour of its users. Our visual analysis of Airbnb's brand evolution, then, contributes to highlighting this paradox, shedding light on the significance of authenticity in brand culture.

CASE STUDY 2: A RECEPTION-LED VISUAL ANALYSIS OF DIGITAL MEDIA AND TECHNOLOGY BRANDS

Much has been written about logos, particularly from the perspectives of graphic designers and scholars with an interest in how different combinations of various shapes, colours, fonts, icons and textures may generate particular connotations and result in more or less attractive logo designs. Research on the formal qualities and meanings of logos has largely been confined to a focus on design and production – that is, to analyses of the semiotic choices or the professional practices, techniques and technologies involved in the creation of logos. Because logos are simple graphics that require a heightened degree of **stylization** and

intentionality, examining their visual resources is especially important in order to understand what their meanings and implications are. As Johannessen (2018: 166–7) explains, 'structural differences' and 'even minute formal qualities' are decisive in constructing or shifting the meanings of a logo.

However, as we know, all visual communication is **polysemic**, or capable of having multiple meanings. One of the key criticisms of semiotics as an approach to visual analysis is that it doesn't take into account the lived experiences and points of view of those who 'make' the meanings of images, either by producing or consuming them. On the other hand, a semiotic approach enables us to examine the key visual traits of images and the subtle differences between them in detail, something that becomes especially important when examining simple graphics like logos.

Research questions and approach

To explore the meanings of logos, then, here we propose a visual analysis based on the views expressed by a group of participants on a selection of logos. In other words, we use our research participants' perspectives to understand how the formal qualities of logos 'work' to make meaning. Our research questions here are:

- What are the key characteristics of 'good' logo design according to a group of research participants? In particular, what are some of the visual resources that make a logo more or less appealing according to these responses?
- How do these responses contribute to an understanding of how logo design 'works' semiotically?

For this case study, we drew inspiration from Floch's famous **semiotic analysis** comparing the Apple and IBM logos (Floch, [1995] 2000). We therefore set out to examine a selection of logos from nine well-known digital media and technology brands, including both IBM and Apple. This focus was also due to the relative ubiquity of social media and online communication services in people's everyday lives, and consequently of their logos. Our case study was designed to elicit and examine responses collected from 91 students who attended the same media and communication class in relation to their attitudes towards these logos. We created a survey centred on an **image-sorting technique** (also known as Q-sort or Q-methodology) paired with **focus groups**, which enabled us to examine participants' subjective attitudes and intuitive interpretations regarding the logos (Lobinger and Brantner, 2015). This approach also enabled us to gather both quantitative and qualitative data about participants' views, in order to understand the reasoning and feelings behind participants' rankings and judgements (O'Neill et al., 2012).

Findings: the visual appeal of simplicity, recognizability and motivation

Our survey was divided into two sections, with Section One asking basic questions about each participant's gender, age and nationality, together with questions about digital media and technology usage. The majority of our participants were between 18 and 24 years of age (98.9%), female (75.8%) and British (63.7%). There were 19 other nationalities represented in the sample, which could be broadly grouped under European (17.6%), Asian (15.3%) and other (4.4%) nationalities. We also found that the majority were primarily Mac users (59.3%), and that the social media platforms that they used most frequently were Facebook and Instagram, respectively, followed by Snapchat and YouTube. Of the two digital communication services included in the survey, WhatsApp was significantly more popular than Skype.

Section Two of the survey was dedicated to our sorting exercise with follow-up questions expanding on participants' views about the logos under evaluation. As a first step here, we asked participants to assign a brand or company name to each of the logos displayed in the survey (Figure 11.7). Most participants recognized all of the logos, but 23 out 91 participants (25%) did not assign a name to the IBM logo, and presumably many others simply filled in 'IBM' without knowing exactly what the logo stood for.

We then asked them to perform two separate sorting tasks. Task A asked participants to sort the logos into three groups, depending on how they felt about each of them and based on their first reaction to each logo. The three groups were 'This is a good logo', 'This logo is neither good nor bad' and 'This is a bad logo'. This task was aimed at eliciting participants' immediate reactions to the logos. There was no limit to how many of the logos could be placed under each category and the wording was kept simple ('good', 'neither good nor bad', 'bad') to allow participants to focus on their immediate reactions to the logo designs rather than on more complex feelings and considerations.

Here we found that the Twitter, Apple and Instagram logos were the three top 'good' logos, with relatively small differences in the percentage of participants who rated each of these logos as 'good'.

FIGURE 11.7 Selection of logos for this case study.

Conversely, the IBM logo was clearly rated as a particularly bad logo by nearly 65% of participants, whereas the second logo in this ranking (i.e. Microsoft) was considered to be a bad logo by less than 19% of our participants (Table 11.1).

TABLE 11.1   Findings from our first sorting exercise (Task A).

This is a good logo	This logo is neither good nor bad	This is a bad logo
Twitter (83.5%)	Microsoft (61.5%)	IBM (64.8%)
Apple (81.3%)	Google (52.7%)	Microsoft (18.7%)
Instagram (72.5%)	Amazon (50.5%)	Snapchat (15.4%)
Facebook (60.4%)	Facebook (40.6%)	Instagram (11%)
Snapchat (52.7%)	Snapchat (31.8%)	Amazon (9.9%)
Google (47.2%)	IBM (29.7%)	Apple (2.2%)
Amazon (42.8%)	Instagram (20.9%)	Facebook (1.1%)
Microsoft (20.9%)	Apple (16.5%)	Google (1.1%)
IBM (7.7%)	Twitter (15.4%)	Twitter (1.1%)

Task B asked participants to order the logos from most appealing to least appealing. Unlike the previous sorting exercise, then, this task entailed a greater degree of reflection as it required participants to place each logo in a different box and to rank the logos in relation to one another. In asking participants to rank logos as more or less 'appealing', we furthermore aimed to elicit their feelings about the logos. When evaluating the logos' appeal, Instagram came in as a clear first, with 40.6% participants considering it the most appealing logo, while more predictably the IBM logo was considered as the least appealing logo (67%). Most notably, the Instagram logo was evaluated as considerably more appealing than the Twitter and Apple logos, both of which had been rated as 'good' by a greater number of participants (Table 11.2).

TABLE 11.2   Findings from our second sorting exercise (Task B).

Most appealing	Neutral (middle)	Least appealing
Instagram (40.6%)	Apple (16.5%)	IBM (67%)
Twitter (27.5%)	Facebook (16.5%)	Microsoft (16.5%)
Google (14.3%)	Google (15.4%)	Amazon (5.5%)
Apple (8.8%)	Snapchat (15.4%)	Snapchat (4.4%)
Facebook (3.3%)	Amazon (11%)	Instagram (3.3%)
Amazon (2.2%)	Twitter (7.7%)	Google (2.2%)
Microsoft (2.2%)	IBM (6.6%)	Apple (1.1%)
Snapchat (1.1%)	Instagram (5.5%)	Facebook (0%)
IBM (0%)	Microsoft (5.5%)	Twitter (0%)

Following these sorting tasks, we asked participants to reflect in writing and in their own words on the reasons why they found particular logos more or less appealing, considering if and how design elements like shape, colour, font, etc. played a role in their evaluation. We also asked them to express their thoughts about the logos of other well-known digital media and technology companies that were not included in the survey and which they found particularly good or bad.

Participants' answers can be divided into three main findings. First, **simplicity** was mentioned by most as the main quality of appealing logos. Overall, this particular group of participants associated simplicity with a logo design's ability to fit neatly in the 'app format'. As one of our participants put it: 'We are used to seeing logos in the shapes of squares, e.g. apps on our phones and devices, therefore the square shape logos are most appealing.' Compact, square shapes, and both smooth lines and corners were privileged signifiers of simplicity and contributed greatly to a logo's appeal.

Second, **recognizability** was another key characteristic of appealing logos. Once again, this characteristic was associated with specific semiotic resources, particularly colour. In speaking about the reasons why she found the Snapchat, Instagram, FB and Twitter logos appealing, for example, one of our participants said: '[T]he colours make it more appealing and stand out. We could compare those to Apple, as it isn't coloured enough and seems a bit boring even though it is meant to look "clean".' Another participant stated: 'I like a logo that is unique and therefore instantly recognizable. I think dull colours often look most professional and trustworthy. However, bright colours can seem friendly and fun.'

A third key characteristic of appealing logos can be defined as **motivation**. Participants thought that logos ought to be clearly linked to the brand's identity and offerings. Hence, there was a strong preference for 'symbols' or 'icons' over wordmarks (or logotypes). The Twitter logo was frequently mentioned as a good example of this clear link between a pictorial logo (or *icotype*; Heilbrunn, 2001) and the brand. As one participant wrote: 'I like how the bird corresponds to the name of the company; this is an interesting word play.' Drawing a comparison with the 'random' Snapchat ghost motif that 'does not quite relate', another participant stated that '[a] bird as a logo for Twitter can be related to the function of "tweeting" and Instagram's camera logo can relate to the function of posting photos.'

Across responses, the Instagram logo seemed to encapsulate all of these three key characteristics of appealing logos. As one of the participants put it: 'I particularly like the Instagram logo because of the nice colours and simplicity, as well as the fact that you can tell that the logo is a camera. It is simple yet effective and aesthetically pleasing.' Participants seemed to be aware that their preference for compact, colourful, pictorial logos was not tied to a judgement of how good a particular logo was. In fact, the Apple logo was evaluated as 'good' by a greater number of participants than the Instagram logo was. On the other hand, the Instagram logo was considered much more appealing than the

Apple logo which, together with the Facebook logo, was considered as most neutral in terms of appeal.

Overall, participants favoured simplicity over other features, granted that this didn't translate into 'boring' designs. Both recognizability and motivation were highlighted as features that made a logo more interesting – for example, through colour and visual puns. Not surprisingly, then, the IBM logo, which was considered to be least appealing, was defined both as fairly simple in its design, but also as 'dull', 'bland', 'basic' and 'outdated'. Many didn't like the black-and-white colour scheme matched with the 'bulky' lettering and striped motif. Along the same lines, participants' thoughts on other logos highlighted a combination of simplicity and recognizability or motivation as key to their appeal. The YouTube logo, in particular, was a favourite, and was described as 'simple and bold', 'neat and clear', 'easily recognizable' and 'bright', but also 'self-explanatory' and 'to the point'. One of the responses summed up the YouTube logo's appeal in the following way: '[I]t's just an icon, yet you directly associate it with the brand along with its very vibrant red colour.'

After the survey, we interviewed ten of our participants in two separate focus groups of six and four participants, respectively. These focus groups enabled participants to elaborate on their own responses to the logos in relation to more general findings from the survey. What emerged here is that logos that were judged as being 'good' thanks to their being 'sleek', 'serious' or 'professional' were not necessarily also the most appealing – this was the case for the Apple and Facebook logos. Focus group participants explained that the Instagram logo was more 'expressive', 'entertaining', 'eccentric' and 'extroverted'. There were also quite a few comments about Instagram's old logo and its rebranding, with a few participants expressing an affective attachment to the old logo which, in the words of one of our participants, 'was quite vintage'. For the same reason, participants also said that they didn't like the new Instagram logo when it was first introduced. On the other hand, as one of our participants put it, the Facebook logo may be seen as 'quite arrogant in its simplicity' because of its iconic use of the letter 'F', which everyone now associates with Facebook, to the extent that no other brand could use the same letter for their logo. Along the same lines, the Apple logo is both 'minimalist' and 'sophisticated' but, by the same token, also 'boring'.

Case study conclusion: complicating a semiotic understanding of logos' visual appeal

Among designers and scholars alike, there seems to be general agreement that the 'simpler' a logo is, the better or more appealing it is (Adams et al., 2004; Mollerup, 1996). According to this view, the notion of 'simplicity' is typically associated with the modernist roots of graphic design, where abstraction is considered as a key method for making purity and truth perceivable

(Cheetham, 1991). By the same token, logo designs centred on figurative motifs (e.g. the bird in the Twitter logo or the siren in the Starbucks logo) or words (e.g. the Coca-Cola and Amazon logos) are often stylized or streamlined. This privileging of simplicity as abstraction and minimalism is related to Marcel Danesi's claim that logos ought to 'bestow upon a brand the timelessness and universality that we associate with primordial mythic narration' (Danesi, 2006: 23). Along the same lines, Floch claims that the Apple logo can be defined as a 'mythogram', because it is able to 'tell' two stories – one of transgression and one of creativity – at the same time, thanks to its visual simplicity and 'non-linear presence' (Floch, [1995] 2000: 53). According to this analysis, the Apple logo is ultimately more appealing than the IBM logo precisely because its 'visual invariants' – e.g. the shape of the bitten apple and the dominance of curves in its contour – work together to make the logo 'simple and strong' (Floch, [1995] 2000: 41).

However, as we saw from our case study, a 'good' logo that looks 'clean' or 'professional' may not be the most appealing. Our survey and focus groups show that the Instagram logo may have been considered as the most appealing precisely because simplicity was matched by recognizability and motivation. Key findings also show that the ways in which participants define general design features depend on their background and on their context. For example, simplicity was associated with visual cues – like compact square shapes and rounded corners – that were tied to a particular logo's ability to fit neatly in the 'app format'. Meanwhile, simplicity was only judged as a positive design feature if a logo was also recognizable and motivated, which in the specific case of our group of participants often meant that logos also ought to be brightly coloured and iconic. These judgements were clearly specific to a demographic that has continuous access to smartphones, is social media savvy, and for whose attention digital media and technology brands compete.

Likewise, participants' feelings about particular logos are also tied to their real-life experiences of brands. Quite a few participants didn't know what IBM was, and it is likely that this lack of previous engagement with the brand also drove some of their negative responses about the logo. The logos of two widely used brands among our participants (Apple and Google) were judged as similarly 'boring' and 'outdated', but their designs were evaluated as being much better than IBM's. Another factor that may very well have led participants to consider the Instagram logo as more appealing than others was its recent redesign, which was vividly recalled by focus group participants. Overall, this kind of **reception-led visual analysis** is important, because it illuminates that seemingly universal principles of 'good' design are also deeply contextual. In combining an investigation of responses by particular groups of research participants with an approach grounded in semiotic analyses of brands and logos like those performed by Floch and Danesi, we can gain a better understanding of how logo design 'works' in people's everyday lives.

CHAPTER CONCLUSION: THE VISUAL POWER OF BRANDING

This chapter has focused on how visual aesthetics are key to defining our everyday relationships with brands. Across the two case studies, we showed how shape and colour, line and typography, texture and finish, and both layout and composition are all visual resources that are central to branding, and all have different communicative functions. There are two main takeaways here:

- There are multiple types of visual resources and media that are available to designers, and their various possible combinations lead to different outcomes in the look and feel of a brand. Changes in the way in which a brand positions itself in relation to potential customers often result from relatively small variations in how the brand is communicated visually, and these variations also have important implications for the brand's economic value.
- Context matters, both in relation to production and reception. The ways in which a brand is perceived often depend on how it is strategically framed by its producers, through media interviews and other promotional means. Conversely, particular groups of users bring different meanings to the table when engaging with widely successful brands.

Overall, visual analysis enables us to identify what brands have to offer that is distinctive and different from other brands, but it also and above all illuminates how brands engender in us particular feelings and attitudes, quite often strategically.

POTENTIAL FURTHER RESEARCH

In this chapter, we have focused on commercial brands, but branding extends far beyond purely commercial realms. Nation branding, in particular, has become integral to governments' policies aiming to attract tourism and foreign investment while also nurturing a sense of belonging and maintaining loyalty among country nationals.

Along the same lines, many of us are familiar with how universities, non-profit organizations and even churches rely on distinctive brand identities to compete for attention and gain traction among publics in otherwise saturated 'marketplaces' like higher education, volunteering and philanthropy, and increasingly also religion.

Politics, too, is a thriving ground for branding. Famously, Barack Obama's success as a presidential candidate in the US elections was largely tied to his highly personalized political brand, which was set apart by values like authenticity, transparency, change and hope. These abstract values were communicated through a combination of promotional media and techniques, including media imagery that portrayed him while playing basketball or mingling with election campaign volunteers, the use of social media to connect directly with potential voters, an election campaign logo in the shape of an 'O' representing the rising sun over the horizon, and Obama's now iconic portrait on the poster designed by the street artist Shepard Fairey.

Some guiding questions in exploring some of these additional areas of brand culture could be: 'What kind of feelings, attitudes, and/or behaviours does a particular nation, organization or politician's brand intend to evoke through visual resources like shape, colour and typography?' 'What do changes in the visual branding of a city, university or political party over time tell us about their social status, economic aspirations and target audiences?'

FURTHER READING

Floch ([1995] 2000) and Danesi (2006) are useful guides for the development of semiotic approaches to logos and other visual dimensions of branding. For visual analyses of specific brands like Nike and Starbucks, see Goldmand and Papson (1998) and Aiello (2018).

12 CONCLUSION

Throughout this book we have presented snapshot visual analyses that make forays into a variety of debates in media and communication studies. We hope to have demonstrated that visual communication scholars have made key contributions to these debates, while also adding our own empirical findings and critical insights to important discussions about the relationship between visual images, the media and our everyday encounters with social, political and cultural issues. And because one of our key aims was to develop a toolkit to examine images across different types of media and through diverse research questions, we also hope to have signalled how our approach could work as a blueprint for analysing alternative and emergent cases.

With regard to this last point, as we moved towards the book's production phase, everyday examples continued to pique our interest and raise significant questions about the role of visual images in media culture. In the November 2018 midterm elections in the US, a record-breaking number of women were elected to the House of Representatives (102 women in a 435-seat House). Alexandria Ocasio-Cortez became the youngest ever woman elected, at 29, and took her seat in January for the Democrats. Being a socialist, young and a woman of colour, Ocasio-Cortez has been a particular focus for hatred from right-wingers who criticize her clothes and question her identity. In an attempt to humiliate her, 'AnonymousQ' posted a video of Ocasio-Cortez from when she was a student, dancing on a roof in a homage to the film *The Breakfast Club*. This cowardly deed spectacularly backfired when many high-profile celebrities and politicians started tweeting their support, and it only seemed to bolster her popularity (along with the music that featured in the video). It's not often that a politician captures people's hearts and wins a social media skirmish. The anonymous poster had failed to think about exactly how this video would be received and the qualities it communicated about Ocasio-Cortez, not only in relation to her political role but also as an appealing (self-)representation of her personal, networked identity.

In January 2019, photographs in *The New York Times* depicting dead bodies of the victims of a terror attack in Nairobi prompted a public outcry, with many Kenyan journalists pointing out that bodies of Western or white victims would not have been shown in such a graphic manner. Not only does this prompt questions about how the media cover terror attacks, but it also speaks to debates about institutionalized racism in Western media reporting and questions about human dignity in depictions of distant suffering. The forms and contents of visual representations continue to play a significant role in questions of ethics and justice. The rapid circulation of images across global media culture often raises critical questions about how audiences view the world and their own capacity to bring about change to the decision processes behind such representations.

Also at the beginning of 2019, the razor company Gillette released an advert titled 'Believe' adapting their well-known slogan 'The best a man can get' to 'The best a man can be'. Directly referencing the #MeToo movement and including clips from news, advertising and other media, the advert depicts men stepping in to prevent sexual harassment, bullying and aggressive behaviour. Perhaps not surprisingly, a backlash followed with angry tweeters calling for a boycott of Gillette products. For a brand known for its almost parodic depictions of handsome manliness, this shift to calling out toxic masculinity has at the very least 'started a conversation'. The primary message of the advert is how more positive portrayals and behaviours can influence the men of the future. Whether deemed a cynical and patronizing ploy for viral publicity or a brave step towards celebrating manhood through positive role models, the visual style of the advert and its break from genre expectations are important aspects to consider in this debate.

These three recent examples cut across the themes of this book – identities, politics, commodities. We find these mediated mini-eruptions fascinating precisely because they reveal the significance of visual images for these overlapping areas of social life together with their imbrication with other major aspects of contemporary media culture. Here, and in the book's 18 case studies, the visual dimension is at the centre of tensions and contestations over meaning, power and credibility – tensions and contestations that arise across different forms of mediation, albeit not always in ways that are directly acknowledged or even self-evident. In the remainder of this short conclusion, we reflect on some of the core concerns for visual communication researchers, once again employing our overarching categories of identities, politics and commodities. We then turn to some of the current issues that we haven't covered in the book, but that we think are intriguing and important areas for research on visual communication in the near future.

IDENTITIES

We have shown how visual images play an important role in how we see ourselves and others. In the current shifts we see across the world, our identities are being shaped and reshaped by media culture, both as part of optimistic movements that seek justice and empowerment like Black Lives Matter or #MeToo and in relation to positionalities that have been thrust upon people due to circumstances beyond their control such as in the case of refugees, but also in a slightly more sinister manner, in populist or reactionary responses to feelings of alienation or lack of control that include, but are not limited to, the *gilets jaunes* in France and the alt-right movement in North America. Social media platforms appear paradoxically to encourage both individualization and the formation of new types of affiliation and collective identities, together with their 'others', across a range of geographical, socio-economic and political vantage points.

The data generated by social media users and how it is put to use have become key concerns for policy-makers, activists and the tech companies themselves.

Our online activities are captured and mapped in invisible ways, raising concerns about how we might be categorized or othered (as rich or poor, haves or have-nots, worthy or unworthy), via mechanisms that remain unseen and largely unknown. We have written about how our identities are performed on social media in the act of sharing both self-created and existing visual content. The act of sharing images, together with liking or tagging the images shared by others, is an important part of our social lives, but the sociability and recognition of others constituted in such image-making and sharing practices leave traces of metadata ripe for behavioural profiling.

Another side to this issue, regarding the ability to access large datasets, concerns journalism, and more specifically data journalism. For example, *The New York Times* software architect, Jake Harris (2015), has reflected on how people are identified in data visualizations and graphics as 'dots' rather than people. Harris writes that such maps and charts can work to remove the human element from the story, and that the graphics need to be balanced through empathic design and reporting that focuses on the stories of the people affected. Both visual communication scholars and practitioners can and ought to ask how their research may promote awareness of the ways in which complex information about people's lives is visually presented, particularly in order to encourage design that contributes to generating greater empathy and understanding (Aiello, 2019). Another ongoing challenge for the scholarly community is to find ways to ensure that the data collected by big tech companies, including images and their metadata, are accessible to a variety of publics, including but not limited to researchers and practitioners (see Niederer, 2018).

POLITICS

We have noted that the manipulation of images for political gain has led to a further erosion of public trust in the nature of information circulated via news websites and social media. In some cases, the image itself is fake, in others, the unmanipulated image is used in misleading and deceptive ways.

While deception and manipulation are by no means new to politics, technological advances in Artificial Intelligence (AI) and machine learning are being employed to create 'deep fakes', or audio and video recordings that portray people doing and saying things they never have. As Robert Chesney and Danielle Citron (2019) explain, deep fakes emerge when pairs of algorithms are pitted against each other in 'generative adversarial networks', or GANs. Here, machine learning algorithms train against each other to continually generate and spot artificial photographic content. Thanks to the availability of a vast database of 'real' images and the ongoing correction of 'fake' content, GANs are able to keep advancing in terms of building plausibility into the creation of artificial photographic images. Chesney and Citron point to the potential for deep fakes to stoke social and ideological divisions when used in political campaigns. As publics become more aware of the possibilities of the technology, this may lead to an even more generalized decline in trust, where not only is seeing no longer believing, but also the very ethics of using audio-visual content as evidence are put into question.

In the context of growing concerns that social media platforms encourage the spread of false news, or more extreme content, these amplification practices require studying in more detail. We therefore ought to interrogate the motivations of those producing and sharing such content, and how the affordances of social media platforms may work to reward such practices. By the same token, visual communication research ought to contribute to upcoming debates on the role of social media companies in the development and implementation of technologies for image verification and authentication, not only in terms of the legal obligations and economic incentives underlying their adoption, but also with regard to their moral and social implications.

Moving on to global politics and conflict, the capacity of photography to 'bear witness' to suffering continues to be debated. For those living in conflict zones or who are displaced through war and insecurity, the question is no longer about the lack of documentary imagery, but also how to cope with the vast digital archives that are often reliant on corporate platforms such as YouTube or Facebook to store them. Changes in policies regarding nudity or graphicness, born of motivations to comply with regulations or protect young users, can lead to the removal of images that serve as witnessing texts to atrocities in Syria and Yemen, as well as more established visual icons of cultural memory, as seen with Facebook's temporary removal of Nick Ut's 'Napalm Girl' photo of Kim Phuc in 2016.

Visual communication research can contribute a deeper understanding of the reasons why certain images rise to the surface in the sea of digital imagery and 'prick' the conscience of audiences, while also engaging with those groups and institutions that have the responsibility to collate, document, classify, protect and ensure accessibility to the disparate images of war and atrocity. Ultimately, this knowledge can also be used to intervene in discussions on how such responsibility might be enforced or governed, and to trace and thus also demystify the networks of actors that amplify certain images over others.

COMMODITIES

We have already noted above the ways in which identities are shaped online via self-representational practices on social media platforms and how the data created by such practices is put to use in opaque ways that stray beyond the transparent performance of our social relationships. But it is not so much the digital technology itself that propels such shifts; rather, the social and economic logics of a new form of capitalism centred on what looks like the provision of free services are, in fact, implicated in treating human experience as material to translate into behavioural data. Applying machine intelligence, 'prediction products' anticipate what users will do or want next. This is what Shoshana Zuboff (2019) has named 'surveillance capitalism', noting that it uses behavioural data based on our sociability online to feed markets about us, not for us. Behaviours come to be actively modified towards desired commercial outcomes, and this is where autonomy becomes eroded. One form of such data is the visual images that are stored, for example, in Facebook or Google Cloud Vision.

Obviously, there is a very well-documented history of how advertisers and marketers have used information about the behaviours and inclinations of different groups of people to target them as consumers and turn them into loyal customers. As we highlighted in the book, however, it is not only products or experiences that are 'sold' in contemporary commerce, but also all-encompassing lifestyles premised upon ethical and political principles. Our social media feeds are rife with explanatory journalism pieces that encourage us to feel and act in particular ways in relation to broader social and cultural phenomena like gender relations or environmentalism, often with significant implications for our everyday purchasing choices. In turn, these pieces are fed to us because of pre-existing information on our consumer behaviours and social networking identities.

Needless to say, both user-created and professionally made visual images are part and parcel of this feedback loop, and we ought to examine their implications for the circulation of lifestyle ideologies as commodities in their own right. And because of the increasingly experiential if not intangible nature of 'goods', visual communication research is key to an understanding of how a variety of visual cues beyond photographic images alone – as, for example, fonts, colour and layout – are also used to shape and profit from the personal and social identities of media users.

THE POWER AND POETICS OF IMAGES

It is hard to escape the role of media technologies and corporations in each of the areas we have outlined here. There are deeply difficult social, ethical and political questions raised in the ways in which humans approach the management and regulation of media infrastructures, software and platforms. If it is true that we are becoming unable to tell apart 'real' and 'fake' images, what does this mean for empathy and understanding of events we only know about through media images? Asko Lehmuskallio, Jukka Häkkinen and Janne Seppänen argue that 'we have to develop a visual literacy that acknowledges the difficulty of separating digital photographs from computer-generated images' (Lehmuskallio et al., 2018: 3).

This is because judgements about the authenticity of photographic images can no longer depend on conventional understandings of photography. Not only has the digitization of photography raised questions about the evidentiary power of photographic images as such, but camera-based and computer-generated images share more features in common than we think. Our ingrained understanding of camera-based images as indexical traces and conversely also of computer-generated images as inauthentic simulations does not take into account the fact that the substance of both types of digital image depends on computational processes. Lehmuskallio et al. (2018) maintain that, in the end, we ought to rely on communities of practice with strong codes of conduct rather than technology alone to both ensure and ascertain their status as records or evidence.

Meanwhile, visual communication texts and practices that we typically would not associate with 'the media' or even with the notion of 'image' have acquired both evidentiary and representational power in media culture. For example, data visualizations are now often used as 'news images' in their own right, particularly in contexts of long-form and data journalism, but also in 'ordinary' news media. As such, data visualizations are also becoming increasingly mundane and standardized, while being largely perceived as objective reflections of social reality (see Anderson, 2018; Engebretsen et al., 2018; Kennedy et al., 2016).

Likewise, digital practices like taking screenshots of online activities and tagging images on social networking platforms are de facto turning both screenshots and social media images into visual records for much broader audiences than originally intended. As Paul Frosh argues, the screenshot has acquired 'the evidentiary power and transparency that have long been ascribed to photography' (2018: 75), and in digital culture it may have even replaced the photograph as the most widely held signifier of authenticity and indexicality. For example, in news media and social media alike, it is now common practice to use screenshots of deleted Twitter posts or web pages to 'prove' that something happened or someone said or did something. In a related, though by no means identical manner, tagging is an act of 'deictic pointing' (Frosh, 2019: 97) that turns the social media images that portray us as individuals into photographs with a referent that can be replicated and circulated in a semi-public manner. In doing this, Frosh argues, the tag anchors our online identities to strong visual evidence of our embodied selves 'through reference, symbol, figure and matter' (Frosh, 2019: 111). On the whole, much of what we do online and which can be produced digitally is not only strikingly visual, but is also susceptible to being 'fixed' as a powerful image or set of images that can then be circulated and used across social networking profiles, digital platforms and media genres – often to support, substantiate or authenticate otherwise fleeting facts and ephemeral identities.

It is also important to emphasize that with each technological evolution, there are anxieties about how our relationship to both reality and imagination is affected. Think of the reports of people running terrified when they first viewed the Lumière Brothers' *Arrival of a Train at the Station* in 1895. As Tom Gunning (1995) has argued, there is a sense of founding myth attached to these reports of early cinema spectatorship, which casts the spectators as childlike, but should instead focus on how the audience are actively delighting in the illusion of the show. Gunning stresses the importance of considering the historical context, in which attending the cinema was an encounter with modernity for those experiencing alienation in an increasingly industrialized and consumer-oriented society. He also reminds the reader of the immediate context – the desire to be thrilled and shocked is stoked in darkened rooms presented as magic theatres.

In the current context, for the majority of media images we consume, there is no single specific built environment – films are watched on the train, adverts surround us on buses and on YouTube videos, smart television sets connect us to the internet and so become conduits for both scheduled television but also networked content, and the 'polymedia' environments enabled by smartphones (Madianou, 2014) make for apparently seamless experiences of different media formats and technological affordances regardless of our location in space. The logics of media convergence and the 'spreadable' (Jenkins et al., 2013) nature of media content come in tandem with new qualitatively different modes of production and reception, in some cases just as disorientating as the technologies and encounters associated with early industrialization and consumer culture noted above.

And the pleasures of discovering new spectacles should not be forgotten in the yearning to fully comprehend our post-internet anxieties. In this regard, we believe that Frosh's original perspective on the poetics of digital media can be more specifically applied to our overarching approach to visual communication. Visual images are not only powerful and power-laden, but they are also poetic insofar as they disclose and produce worlds 'in which we find and make our meaning' (Frosh, 2019: 33). It is in this sense that we see visual images not only in terms of their problems but also their potentials – for moral engagement, creative disruption, and both cultural and social change. We have already noted the cross-cutting nature of our primary organizing themes of identities, politics and commodities. Behind each of our enquiries into visual communication, then, is a deep interest in how images are implicated in the construction of our social and political worlds.

Whether interrogating the sites of image-making practices, media platforms and genres, or modes of spectatorship and consumption, our approach is underpinned by an interest in the interplay of media culture and public life. We follow Robert Hariman and John Lucaites here in their statement that '[i]mages, whether statues in antiquity or photographs today, are an essential means for creating a public world' (2016: 19). This statement moves beyond the question of how images are 'made public', to emphasize their performative nature in constituting the space necessary for citizenship. While the authors are primarily writing about photojournalism, we agree that it would be productive to move away from the premise that the polysemic nature of visual communication is a problem, towards one that embraces the plurality and openness of images, and values the diverse ways in which they act as 'a means for communicating with others about common concerns' (Hariman and Lucaites, 2016: 5).

This also means continuing to draw upon a range of theories, methods and disciplinary knowledge in order to better understand the meaning-making capacities of images through rich, interpretive encounters, rather than proposing a single method of enquiry. In this continuing endeavour, we echo Roland Bleiker's (2018) call for pluralist approaches to visual global politics, a sensibility we would extend to visual artefacts, visibilities and visualities not necessarily deemed political in nature. By the same token, in this book we have privileged and promoted an overarching critical outlook centred on a close analytical focus on visual images themselves, in combination with considerations about key aspects of production, circulation and audiencing. It is precisely by going back to images in their own right, we believe, that we can understand both their power and their poetics. As a whole, we hope that our book has illuminated at least some of the ways in which images make sense, make meaning, make worlds and ultimately also make things happen in contemporary media culture.

REFERENCES

Abidin, C. (2018) *Internet Celebrity: Understanding Fame Online*. Bingley: Emerald.

Acquah, N.K., DiCampo, P., Merrill, A. and van der Heijden, T. (eds) (2017) *Everyday Africa: 30 Photographers Repicturing a Continent*. Heidelberg: Kehrer Verlag.

Adami, E. (2015) Aesthetics and Identity in Digital Texts Beyond Writing: A Social Semiotic Multimodal Framework. In Archer, A. and Breuer, E. (eds) *Multimodality in Writing: The State of the Art in Theory, Methodology and Pedagogy*. Leiden: Brill, pp. 43–62.

Adams, S., Morioka N. and Stone, T. (2004) *Logo Design Workbook: A Hands-On Guide to Creating Logos*. Beverly, MA: Rockford Publishers.

Adorno, T. and Horkheimer, M. ([1944] 1993) The Culture Industry: Enlightenment as Mass Deception. In Adorno, T. and Horkheimer, M. (eds) *Dialectic of Enlightenment*. New York: Continuum.

Aiello, G. (2007) The Appearance of Diversity: Visual Design and the Public Communication of EU Identity. In Bain, J. and Holland, M. (eds) *European Union Identity: Perceptions from Asia and Europe*. Baden-Baden: Nomos, pp. 147–81.

Aiello, G. (2012a) The 'Other' Europeans: The Semiotic Imperative of Style in *Euro Visions* by Magnum Photos. *Visual Communication*, 11(1): 49–77.

Aiello, G. (2012b) All Tögethé® Now: The Recontextualization of Branding and the Stylization of Diversity in EU Public Communication. *Social Semiotics*, 22(4): 459–86.

Aiello, G. (2016) Taking Stock. *Ethnography Matters*. Available at: https://ethnographymatters.net/blog/2016/04/28/taking-stock/ (accessed 12 March 2019).

Aiello, G. (2018) Losing to Gain: Balancing Style and Texture in the Starbucks Logo. In Johannessen, C.M. and van Leeuwen, T. (eds) *The Materiality of Writing: A Trace Making Perspective*. London: Routledge, pp. 195–210.

Aiello, G. (2019) Inventorizing, Situating, Transforming: Social Semiotics and Data Visualization. In Kennedy, H. and Engebretsen, M. (eds) *Data Visualization in Society*. Amsterdam: University of Amsterdam Press.

Aiello, G. (2020) Visual Semiotics: Key Concepts and New Directions. In Pauwels, L. and Mannay, D. (eds) *The SAGE Handbook of Visual Research Methods* (2nd edn). London: SAGE.

Aiello, G. and Dickinson, G. (2014) Beyond Authenticity: A Visual-Material Analysis of Locality in the Global Redesign of Starbucks Stores. *Visual Communication*, 13(3): 303–21.

Aiello, G. and Parry, K. (2015) Aesthetics, Political. In Mazzoleni, G. (ed.) *International Encyclopedia of Political Communication*. Hoboken, NJ: Wiley Blackwell, pp. 11–16.

Aiello, G. and Pauwels, L. (2014) Special Issue: Difference and Globalization. *Visual Communication*, 13(3): 275–85.

Aiello, G. and Thurlow, C. (2006) Symbolic Capitals: Visual Discourse and Intercultural Exchange in the European Capital of Culture Scheme. *Language and Intercultural Communication*, 6(2): 148–62.

Aiello, G. and Woodhouse, A. (2016) When Corporations Come to Define the Visual Politics of Gender: The Case of Getty Images. *Journal of Language and Politics*, 15(3): 351–66.

Airbnb (2011) Airbnb Free Photography: Celebrating 13,000 Verified Properties & Worldwide Launch. Available at: https://blog.atairbnb.com/airbnb-photography-celebrating-13000-verified/ (accessed 10 March 2019).

Akbarzadeh, S. (2016) The Muslim Question in Australia: Islamophobia and Muslim Alienation. *Journal of Muslim Minority Affairs*, 36(3): 323–33.

Alsultany, E. (2012) *Arabs and Muslims in the Media: Race and Representation After 9/11*. New York: New York University Press.

Anderson, B. (1983) *Imagined Communities: Reflections on the Origins and Spread of Nationalism*. London: Verso.

Anderson, C.W. (2018) *Apostles of Certainty: Data Journalism and the Politics of Doubt*. Oxford: Oxford University Press.

Arnheim, R. (1947) Perceptual Abstraction and Art. *Psychological Review*, 54(2): 66–82.

Aronczyk, M. (2013) *Branding the Nation: The Global Business of National Identity*. Oxford: Oxford University Press.

Arpan L.M., Baker, K., Lee, Y., Jung, T., Lorusso, L., and Smith, J. (2006) News Coverage of Social Protests and the Effects of Photographs and Prior Attitudes. *Mass Communication & Society* 9(1): 1–20.

Arvidsson, A. (2006) *Brands: Meaning and Value in Media Culture*. London and New York: Routledge.

Askanius, T. (2010) Video Activism 2.0 – Space, Place and Audiovisual Imagery. In Hedling, E., Hedling, O. and Jönsson, M. (eds) *Regional Aesthetics: Locating Swedish Media*. Stockholm: Kungliga Biblioteket, pp. 337–58.

Bal, M. and Bryson, N. (1991) Semiotics and Art History. *The Art Bulletin*, 73(2): 174–208.

Banet-Weiser, S. (2012) *Authentic™: The Politics of Ambivalence in a Brand Culture*. New York: New York University Press.

Barnhurst, K.G. and Quinn, K. (2012) Political Visions: Visual Studies in Political Communication. In Semetko, H.A. and Scammell, M. (eds) *The SAGE Handbook of Political Communication*. London: SAGE, pp. 276–91.

Barnhurst, K.G., Vari, M. and Rodríguez, Í. (2004) Mapping Visual Studies in Communication. *Journal of Communication*, 54(4): 616–44.

Barthes, R. (1972) *Mythologies*. New York: The Noonday Press.

Barthes, R. ([1961–73] 1977) Rhetoric of the Image. In *Image, Music, Text*. New York: Hill & Wang, pp. 32–51.

Barthes, R. ([1964] 1977) *Elements of Semiology*. New York: Hill & Wang.

Barthes, R. (1979) *The Eiffel Tower and Other Mythologies*. Berkeley, CA: University of California Press.

Barthes, R. (1982) *Camera Lucida: Reflections on Photography*. London: Cape.

Batziou, A. (2015) A Christmas Tree in Flames and Other – Visual – Stories: Looking at the Photojournalistic Coverage of the Greek Protests of December 2008. *Social Movement Studies*, 14(1): 22–41.

Bednarek, M. (2014) And They All Look Just the Same? A Quantitative Survey of Television Title Sequences. *Visual Communication*, 13(2): 125–45.

Bell, D. and Hollows, J. (eds) (2005) *Ordinary Lifestyles: Popular Media, Consumption and Taste*. Maidenhead: Open University Press.

Bell, P. (2001) Content Analysis of Visual Images. In van Leeuwen, T. and Jewitt, C. (eds) *The SAGE Handbook of Visual Analysis*. London: SAGE, pp. 10–34.

Bell, P. and Milic, M. (2002) Goffman's Gender Advertisements Revisited: Combining Content Analysis with Semiotic Analysis. *Visual Communication*, 1(2): 203–22.

Benjamin, W. (1968) *Illuminations*. New York: Schocken Books.

Bennett, W.L. and Segerberg, A. (2012) The Logic of Connective Action. *Information, Communication & Society*, 15(5): 739–68.

Bennetts, M. (2014) *Kicking the Kremlin: Russia's New Dissidents and the Battle to Topple Putin*. London: Oneworld.

Berelson, B. ([1952] 1971) *Content Analysis in Communication Research*. New York: Hafner.

Berger, A.A. (1998) *Seeing is Believing: An Introduction to Visual Communication*. Mountain View, CA: Mayfield.

Berger, J. (1972) *Ways of Seeing*. London: British Broadcasting Corporation and Penguin Books.

Berger, J. (1980) *About Looking*. New York: Vintage.

Bignell, J. (2002) *Media Semiotics*. Manchester: Manchester University Press.

Billig, M. (1995) *Banal Nationalism*. London: SAGE.

Birdsell, D.S. and Groarke, L. (1996) Toward a Theory of Visual Argument. *Argumentation and Advocacy*, 33(1): 1–10.

Bleiker, R. (ed.) (2018) *Visual Global Politics*. Oxford: Routledge.

Borderless City: European Capital of Culture–Pécs, 2010 (n.d.) Pécs, Hungary: Pécs Application Centre, Europe Centre Pbc.

Bordwell, D. and Thompson, K. ([2004] 2017) *Film Art: An Introduction*. New York: McGraw-Hill Education.

Borger, J. (2016) French Media to Stop Publishing Photos and Names of Terrorists. *The Guardian*. Available at: www.theguardian.com/media/2016/jul/27/french-media-to-stop-publishing-photos-and-names-of-terrorists (accessed 11 January 2019).

Boudana, S., Frosh, P. and Cohen, A. (2017) Reviving Icons to Death: When Historic Photographs Become Digital Memes. *Media, Culture & Society*, 39(8): 1210–30.

Bounegru, L., Gray, J., Venturini, T. et al. (2017) A Field Guide to 'Fake News' and Other Information Disorders. Amsterdam: Public Data Lab with support from First Draft.

boyd, d. (2014) *It's Complicated: The Social Lives of Networked Teens*. New Haven, CT: Yale University Press.

Brideau, K. and Berret, C. (2014) A Brief Introduction to Impact: 'The Meme Font'. *Journal of Visual Culture*, 13(3): 307–13.

Brown, S. and Ponsonby-McCabe, S. (eds) (2014) *Brand Mascots: And Other Marketing Animals*. London: Routledge.

Bruun Vaage, M. (2015) *The Antihero in American Television*. London: Routledge.

Burgen, S. (2017) Barcelona Cracks Down on Airbnb Rentals with Illegal Apartment Squads. *The Guardian*. Available at: www.theguardian.com/technology/2017/jun/02/airbnb-faces-crackdown-on-illegal-apartment-rentals-in-barcelona (accessed 10 March 2019).

Burgess, J. and Green, J. (2018) *YouTube: Online Video and Participatory Culture*. Cambridge: Polity.

Burke, K. (1952) *A Grammar of Motives*. New York: Prentice Hall.

Butler, J. (1990) *Gender Trouble*. New York: Routledge.

Cain Miller, C. (2018). What Being Transgender Looks Like, According to Stock Photography. *The New York Times*, 25 October. Available at: www.nytimes.com/2018/10/25/upshot/what-being-transgender-looks-like-according-to-stock-photography.html (accessed 13 March 2019).

Carr, A. (2014a) Airbnb Reveals a Major Rebranding Effort that Paves the Way for Sharing More than Homes. *Fast Company*. Available at: www.fastcompany.com/3033130/airbnb-unveils-a-major-rebranding-effort-that-paves-the-way-for-sh (accessed 13 March 2019).

Carr, D. (2014b) Its Edge Intact, Vice is Chasing Hard News. *The New York Times*, 24 August. Available at: www.nytimes.com/2014/08/25/business/media/its-edge-intact-vice-is-chasing-hard-news-.html?_r=0 (accessed 13 March 2019).

Chan, J.M. and Lee, C.-C. (1984) The Journalistic Paradigm on Civil Protests: A Case Study of Hong Kong. In Arno, A. and Dissanayake, W. (eds) *The News Media in National and International Conflict*. Boulder: Westview, pp. 183–202.

Chandler, D. (2007) *Semiotics: The Basics*. London and New York: Routledge.

Cheetham, M.A. (1991) *The Rhetoric of Purity: Essentialist Theory and the Advent of Abstract Painting*. Cambridge: Cambridge University Press.

Chen, A. and Machin, D. (2014). The Local and the Global in the Visual Design of a Chinese Women's Lifestyle Magazine: A Multimodal Critical Discourse Approach. *Visual Communication*, 13(3): 287–301.

Chesney, R. and Citron, D. (2019) Deepfakes and the New Disinformation War: The Coming Age of Post-Truth Geopolitics. *Foreign Affairs*, January/February issue. Available at: www.foreignaffairs.com/articles/world/2018-12-11/deepfakes-and-new-disinformation-war (accessed 13 March 2019).

Chouliaraki, L. (2006) *The Spectatorship of Suffering*, London: SAGE.

Chouliaraki, L. and Stolić, T. (2017) Rethinking Media Responsibility in the Refugee 'Crisis': A Visual Typology of European News. *Media, Culture & Society*, 39(8): 1162–77.

Christian, A.J. (2009) Real Vlogs: The Rules and Meanings of Online Personal Videos. *First Monday*, 14(11). Available at: https://firstmonday.org/ojs/index.php/fm/article/view/2699/2353 (accessed 13 March 2019).

Clair, J. (1998) *Henri Cartier-Bresson: Europeans*. Boston, MA: Bulfinch.

Clarke, G. (1992) Introduction. In Clarke, G. (ed.) *The Portrait in Photography*. London: Reaktion Books, pp. 1–5.

Cohen, A.A., Boudana, S. and Frosh, P. (2018) You Must Remember This: Iconic News Photographs and Collective Memory. *Journal of Communication*, 68(3): 453–79.

Commission of the European Communities (2005) Proposal for a Decision of the European Parliament and of the Council establishing a community action for the European Capital of Culture Event for the years 2007 to 2019, 30 May. COM(2005) 209 final, 2005/102 (COD). Available at: http://eur-lex.europa.eu/LexUriServ/LexUriServ.do?uri=CELEX: 52005PC0209:EN:HTML (accessed 10 March 2019).

Corner, J. (2000) Mediated Persona and Political Culture: Dimensions of Structure and Process. *European Journal of Cultural Studies*, 3(3): 386–402.

Corner, J. and Pels, D. (2003) *Media and the Restyling of Politics: Consumerism, Celebrity and Cynicism*. London: SAGE.

Corrigall-Brown, C. and Wilkes, R. (2012) Picturing Protest: The Visual Framing of Collective Action by First Nations in Canada. *American Behavioral Scientist*, 56(2): 223–43.

Cottle, S. and Lester, L. (eds) (2011) *Transnational Protests and the Media*. New York: Peter Lang.

Dan, V. and Iorgoveanu, A. (2013) Still On the Beaten Path: How Gender Impacted the Coverage of Male and Female Romanian Candidates for European Office. *The International Journal of Press/Politics*, 18(2): 208–33.

Danesi, M. (2006) *Brands*. New York: Routledge.

Daniels, J. (2009) Cloaked Websites: Propaganda, Cyber-Racism and Epistemology in the Digital Era. *New Media & Society*, 11(5): 659–83.

Davis, A. (2013) *Promotional Cultures: The Rise and Spread of Advertising, Public Relations, Marketing and Branding*. Cambridge: Polity.

Dean, J. (2018) Sorted for Memes and Gifs: Visual Media and Everyday Digital Politics. *Political Studies Review*, Online First.

DeLuca, K. (1999) *Image Politics: The New Rhetoric of Environmental Activism*. New York and London: Guilford Press.

DeLuca, K. and Peeples, J. (2002) From Public Sphere to Public Screen: Democracy, Activism, and the "Violence" of Seattle. *Critical Studies in Media Communication*, 19(2): 125–51.

Dencik, L. and Allan S. (2017) In/visible Conflicts: NGOs and the Visual Politics of Humanitarian Photography. *Media, Culture & Society*, 39(8): 1178–93.

Der Weg zum Logo Linz 2009 [The way towards Linz 2009's logo] (2006) (Press release, 13 June) Available at: www.linz09.at/en/detailseite/programm/informationen06/ presse-information06/755305.html (accessed 13 March 2019).

de Saussure, F. ([1916] 1983) *Course in General Linguistics*. London: Duckworth.

DiCampo, P. (2012) Ivory Coast: Everyday Africa. *Pulitzer Center*, 15 May. Available at: https://pulitzercenter.org/reporting/ivory-coast-everyday-africa#slideshow-0 (accessed 12 January 2019).

Djonov, E. and van Leeuwen, T. (2011) The Semiotics of Texture: From Tactile to Visual. *Visual Communication*, 10(4): 541–64.

Doerr, N., Mattoni, A. and Teune, S. (eds) (2013) *Advances in the Visual Analysis of Social Movements*. Research in Social Movements, Conflicts and Change, Volume 35. Bingley: Emerald.

Doerr, N., Mattoni, A. and Teune, S. (2014) Visuals in Social Movements. In della Porta, D. and Diani, M. (eds) *The Oxford Handbook of Social Movements*. Oxford: Oxford University Press, pp. 557–66.

Dotschkal, J. (2014) Life in Africa, Unfiltered. *National Geographic*, 5 June. Available at: www.nationalgeographic.com/photography/proof/2014/06/05/life-in-africa-unfiltered/ (accessed 12 January 2019).

du Gay, P., Hall, S., Janes, L., Mackay, H. and Negus, K. (1997) *Doing Cultural Studies: The Story of the Sony Walkman*. London: SAGE with The Open University.

Duguay, S. (2016) Lesbian, Gay, Bisexual, Trans, and Queer Visibility Through Selfies: Comparing Platform Mediators Across Ruby Rose's Instagram and Vine Presence. *Social Media + Society*, 2(2): 1–12.

Dunne, C. (2013) Move Over, Dove. "Orange Is The New Black" Celebrates Real Women. *Fast Company*. Available at: www.fastcompany.com/1673132/move-over-dove-orange-is-the-new-black-celebrates-real-women#1 (accessed 2 September 2018).

Eco, U. (1984) *Semiotics and the Philosophy of Language*. London: Macmillan.

Edwards, J.L. and Winkler, C.K. (1997) Representative Form and the Visual Ideograph: The Iwo Jima Image in Editorial Cartoons. *Quarterly Journal of Speech*, 83(3): 289–310.

Engebretsen, M., Kennedy, H. and Weber, W. (2018) Data Visualization in Scandinavian Newsrooms: Emerging Trends in Journalistic Visualization Practices. *Nordicom Review*, 39(2): 3–18.

Entman, R.M. (1993) Toward Clarification of a Fractured Paradigm. *Journal of Communication*, 43(4): 51–8.

Estrin, J. (2012) Picturing Everyday Life in Africa. 17 September. Available at: https://lens.blogs.nytimes.com/2012/09/17/picturing-everyday-life-in-africa/ (accessed 14 January 2019).

European Capital of Culture (n.d.) Guide for Cities Applying for the Title of European Capital of Culture. Available at: http://ec.europa.eu/culture/eac/ecocs/pdf_word/guide_to_candidates_en.pdf (accessed 13 March 2008).

European Parliament and Council of the European Union (2006, November 3). Decision No 1622/2006/EC of the European Parliament and of the Council of 24 October 2006 establishing a Community action for the European Capital of Culture event for the years 2007 to 2019. *Official Journal of the European Union L 304/1-6*. Available at: https://eur-lex.europa.eu/legal-content/EN/TXT/?uri=uriserv:OJ.L_.2006.304.01.0001.01.ENG&toc=OJ:L:2006:304:TOC (accessed July 10 2019).

Evan, N. (2013) O2 'Be More Dog' Advert is YouTube Hit as Canine-like Cat Becomes Viral Sensation. *Daily Mirror*. Available at: www.mirror.co.uk/news/weird-news/o2-be-more-dog-advert-2032166 (accessed 6 August 2018).

Evans I. (2016) 'I wasn't Afraid. I Took a Stand in Baton Rouge Because Enough is Enough'. *The Guardian*. Available at: www.theguardian.com/commentisfree/2016/jul/22/i-wasnt-afraid-i-took-a-stand-in-baton-rouge-because-enough-is-enough (accessed 10 January 2019).

Fahmy, S. (2007) 'They Took it Down': Exploring Determinants of Visual Reporting in the Toppling of the Saddam Statue in National and International Newspapers. *Mass Communication & Society*, 10(2): 143–70.

Fahmy S., Bock M.A. and Wanta, W. (2014) *Visual Communication Theory and Research: A Mass Communication Perspective*. New York: Palgrave Macmillan.

Fairclough, N. (1995a) *Critical Discourse Analysis: The Critical Study of Language*. London: Longman.

Fairclough, N. (1995b) *Media Discourse*. London: Arnold.

Fairclough, N. (2002) Language in New Capitalism. *Discourse & Society*, 13(2): 163–6.

Farkas, J., Schou, J. and Neumayer, C. (2017) Cloaked Facebook Pages: Exploring Fake Islamist Propaganda in Social Media. *New Media & Society*, 20(5): 1850–67.

Farr, B. (2016) Seeing Blackness in Prison: Understanding Prison Diversity on Netflix's Orange is the New Black. In McDonald, K. and Smith-Rowsey, D. (eds) *The Netflix Effect: Technology and Entertainment in the 21st Century*. London: Bloomsbury, pp. 155–69.

Featherstone, M. (1987) Lifestyle and Consumer Culture. *Theory, Culture & Society*, 4(1): 55–70.

Finnegan, C.A. and Kang, J. (2004) "Sighting" the Public: Iconoclasm and Public Sphere Theory. *Quarterly Journal of Speech*, 90(4): 377–402.

Fitzpatrick, A. (2018) It's Not Just Black Panther: Afrofuturism Is Having a Moment. *Time*. Available at: http://time.com/5246675/black-panther-afrofuturism/ (accessed 13 March 2019).

Floch, J.-M. (1985) *Petites Mythologies de l'Oeil et de l'Esprit: Pour une Sémiotique Plastique*. Paris-Amsterdam: Hadès-Benjamins.

Floch, J.-M. ([1995] 2000) *Visual Identities*. New York: Continuum.

Fornäs, J. (2012) *Signifying Europe*. Bristol: Intellect.

Forrest, D. and Johnson, B. (eds) (2017) *Social Class and Television Drama in Contemporary Britain*. London: Palgrave Macmillan.

Foss, S.K. (2005) Theory of Visual Rhetoric. In Smith, K.L., Moriarty, S., Barbatsis, G. and Kenney, K. (eds) *Handbook of Visual Communication: Theory, Methods, and Media*. New York: Routledge, pp. 141–52.

Foucault, M. ([1975] 1995). *Discipline and Punish: The Birth of the Prison*. New York: Vintage.

Freelon, D., McIlwain, C.D. and Clark, M.D. (2016) Beyond the Hashtags: #Ferguson, #Blacklivesmatter, and the Online Struggle for Offline Justice. *Centre for Media and Social Impact*. Available at: http://cmsimpact.org/resource/beyond-hashtags-ferguson-blacklivesmatter-online-struggle-offline-justice/ (accessed 17 June 2018).

Frosh, P. (2003) *The Image Factory: Consumer Culture, Photography and the Visual Content Industry*. Oxford: Berg.

Frosh, P. (2019) *The Poetics of Digital Media*. Cambridge: Polity.

Garfield, S. (2010) *Just My Type: A Book About Fonts*. London: Profile Books.

Geise, S. and Baden, C. (2015) Putting the Image Back into the Frame: Modeling the Linkage between Visual Communication and Frame-Processing Theory. *Communication Theory*, 25(1): 46–69.

Gerbaudo, P. (2015) Protest Avatars as Memetic Signifiers: Political Profile Pictures and the Construction of Collective Identity on Social Media in the 2011 Protest Wave. *Information, Communication & Society*, 18(8): 916–29.

Gessen, M. (2014) Pussy Riot: Behind the Balaclavas. *The Guardian*. Available at: https://www.theguardian.com/books/2014/jan/24/pussy-riot-behind-balaclava (accessed 18 June 2018).

Giddens, A. (1991) *Modernity and Self-Identity*. Stanford, CA: Stanford University Press.

Gill, R. (2008) Empowerment/Sexism: Figuring Female Sexual Agency in Contemporary Advertising. *Feminism & Psychology*, 18(1): 35–60.

Gitlin, T. (1980) *The Whole World is Watching: Mass Media in the Making and Unmaking of the New Left*. Berkeley, CA: University of California.

Goffman, E. (1959) *The Presentation of Self in Everyday Life*. New York: Penguin Books.

Goffman, E. (1976) *Gender Advertisements*. New York: Harper & Row.

Goldman, R. and Papson, S. (1998) *Nike Culture: The Sign of the Swoosh*. London: SAGE.

Gombrich, E.H. (1960) *Art and Illusion: A Study in the Psychology of Pictorial Representation*. London: Phaidon.

Goodnow, T. (2006) On Black Panthers, Blue Ribbons, & Peace Signs: The Function of Symbols in Social Campaigns. *Visual Communication Quarterly*, 13(3): 166–79.

Grabe, M.E. and Bucy E.P. (2009) *Image Bite Politics: News and the Visual Framing of Elections*. New York: Oxford University Press.

Gregg, M. (2004) A Mundane Voice. *Cultural Studies*, 18(2–3): 363–83.

Greimas, A.J., Collins, F. and Perron, P. (1989) Figurative Semiotics and the Semiotics of the Plastic Arts. *New Literary History*, 20(3): 627–49.

Grossman, P. (2015) Leaning In Looking Back, Forging Ahead – The Journey to Repicture Women. Visual Insights, Stories & Trends. Getty Images, 14 February. Available at: http://stories.gettyimages.com/leaning-looking-back-forging-ahead-journey-repicture-women/1 (accessed 10 September 2018).

Gunning, T. (1995) An Aesthetics of Astonishment: Early Film and the (In)Credulous Spectator. In Williams, L. (ed.) *Viewing Positions: Ways of Seeing Film*. New Brunswick, NJ: Rutgers University Press, pp. 114–33.

Hall, S. ([1973] 1980) Encoding/decoding. In Studies CfCC (ed.) *Culture, Media, Language: Working Papers in Cultural Studies, 1972–79* (2nd edn). London: Hutchinson, pp. 128–38.

Hall, S. (1982) The Rediscovery of 'Ideology': Return of the Repressed in Media Studies. In Bennett, T., Curran, J., Gurevitch, M. and Wollacott, J. (eds) *Culture, Society and the Media*. London: Methuen, pp. 56–90.

Hall, S. (1996) Who Needs 'Identity'? In Hall, S. and Du Gay, P. (eds), *Questions of Cultural Identity*. London: SAGE, pp. 15–30.

Hall, S. (ed.) (1997) *Representation: Cultural Representations and Signifying Practices*. London: SAGE with The Open University.

Hall, S. (2003) 'In but not of Europe': Europe and its Myths. In Passerini, L. (ed.) *Figures d'Europe: Images and Myths of Europe*. Brussels: Peter Lang, pp. 35–46.

Hall, S. (2017) *The Fateful Triangle: Race, Ethnicity, Nation*. Cambridge, MA: Harvard University Press.

Halliday, M.A.K. (1978) *Language as Social Semiotic*. London: Arnold.

Halliday, M.A.K. (1985) *An Introduction to Functional Grammar*. London: Arnold.

Hamel, K.J. (2012) Teaching with Trailers: The Pedagogical Value of Previews for Introducing Film Analysis. *Journal of Film and Video*, 64: 38–49.

Hamilton, P. (1997) Representing the Social: France and Frenchness in Post-War Humanist Photography. In Hall, S. (ed.) *Representation: Cultural Representations and Signifying Practices*. London: SAGE, pp. 75–150.

Hansen, A. and Machin, D. (2019) *Media and Communication Research Methods*. London: Red Globe Press.

Hariman, R. and Lucaites, J. (2007a) *No Caption Needed: Iconic Photographs, Public Culture, and Liberal Democracy*. Chicago: University of Chicago Press.

Hariman, R. and Lucaites, J. (2007b) The Times Square Kiss: Iconic Photography and Civic Renewal in U.S. Public Culture. *The Journal of American History*, 94(1): 122–31.

Hariman, R. and Lucaites, J. (2016) *The Public Image: Photography and Civic Spectatorship*. Chicago: University of Chicago Press.

Harmer, E., Savigny, H. and Ward, O. (2017) 'Are You Tough Enough?' Performing Gender in the UK Leadership Debates 2015. *Media, Culture & Society*, 39(7): 960–75.

Harris, J. (2015) Connecting with the Dots. Source Open News, 15 January. Available at: https://source.opennews.org/articles/connecting-dots/ (accessed 13 March 2019).

Hearn, A. (2008) 'Meat, Mask, Burden': Probing the Contours of the Branded 'Self'. *Journal of Consumer Culture*, 8(2): 197–217.

Heilbrunn, B. (2001) *Le Logo*. Paris: Presses Universitaires de France.

Highfield, T. and Leaver, T. (2016) Instagrammatics and Digital Methods: Studying Visual Social Media, from Selfies and GIFs to Memes and Emoji. *Communication Research and Practice*, 2(1): 47–62.

Hill, C.A. and Helmers, M. (2004) *Defining Visual Rhetorics*. Mahwah, NJ: Lawrence Erlbaum.

Hobsbawm, E. (1983) Mass-producing Traditions: Europe, 1870–1914. In Hobsbawm, E. and Ranger, T. (eds) *The Invention of Tradition*. Cambridge: Cambridge University Press, pp. 263–307.

Holland, J. and Wright, K.A.M. (2017) The Double Delegitimatisation of Julia Gillard: Gender, the Media, and Australian Political Culture. *Australian Journal of Politics & History*, 63(4): 588–602.

Holm, N. (2017) *Advertising and Consumer Society: A Critical Introduction*. London: Palgrave Macmillan.

Horak, L. (2014) Trans on YouTube: Intimacy, Visibility, Temporality. *TSQ: Transgender Studies Quarterly*, 1(4): 572–85.

HPSCI (2017) *HPSCI Minority Open Hearing Exhibits*. Available at: https://democrats-intelligence.house.gov/hpsci-11-1/hpsci-minority-open-hearing-exhibits.htm (accessed 10 March 2019).

Humprecht, E. and Esser, F. (2017) A Glass Ceiling in the Online Age? Explaining the Underrepresentation of Women in Online Political News. *European Journal of Communication*, 32(5): 439–56.

Iedema, R. (2003) Multimodality, Resemiotization: Extending the Analysis of Discourse as Multi-semiotic Practice. *Visual Communication*, 2(1): 29–57.

Iyengar, R. (2015) Protesters Occupy Airbnb Headquarters on Eve of San Francisco Vote. *Time*. Available at: http://time.com/4097686/airbnb-protest-headquarters-san-francisco-vote/ (accessed 10 January 2019).

Jack, L. (2014) How Airbnb Plans to Become a Community-driven Superbrand. *Fast Company*. Available at: www.fastcompany.com/3050679/how-airbnb-plans-to-become-a-community-driven-superbrand (accessed 10 January 2010).

Jackson, S.J. and Foucault Welles, B. (2016) #Ferguson is Everywhere: Initiators in Emerging Counterpublic Networks. *Information, Communication & Society* 19(3): 397–418.

Jacobs, S. (2016) Instagramming Africa. *Journal of African Media Studies*, 8(1): 91–102.

Jenkins, H., Ford, S. and Green, J. (2013) *Spreadable Media: Creating Value and Meaning in a Networked Culture*. New York: New York University Press.

Johannessen, C.M. (2017) Experiential Meaning Potential in the Topaz Energy Logo: A Framework for Graphemic and Graphetic Analysis of Graphic Logo Design. *Social Semiotics*, 27(1): 1–20.

Johannessen, C.M. (2018) The Challenge of Simple Graphics for Multimodal Studies: Articulation and Time Scales in Fuel Retail Logos. *Visual Communication*, 17(2): 163–85.

Johannessen, C.M. and van Leeuwen, T. (2018). (Ir)regularity. In Johannessen, C.M. and van Leeuwen, T. (eds) *The Materiality of Writing: A Trace Making Perspective*. London: Routledge, pp. 175–91.

Jones, J. and Frizzell, N. (2016) The Baton Rouge Protester: 'A Botticelli Nymph Attacked by Star Wars Baddies'. *The Guardian*. Available at: www.theguardian.com/artanddesign/2016/jul/12/baton-rouge-protester-botticelli-nymph-attacked-by-star-wars-baddies-iesha-evans (accessed 10 January 2019).

Kaneva, N. and Popescu, D. (2011) National Identity Lite: Nation Branding in Post-Communist Romania and Bulgaria. *International Journal of Cultural Studies*, 14(2): 191–207.

Kennedy, H., Hill, R., Aiello, G. and Allen, W. (2016) The Work that Visualisation Conventions Do. *Information, Communication and Society*, 19(6): 715–35.

Kernan, L. (2004) *Coming Attractions: Reading American Movie Trailers*. Austin, TX: University of Texas Press.

Kharroub, T. and Bas, O. (2016) Social Media and Protests: An Examination of Twitter Images of the 2011 Egyptian Revolution. *New Media & Society*, 18(9): 1973–92.

Khatib, L. (2013) *Image Politics in the Middle East: The Role of the Visual in Political Struggle*. London: I.B. Tauris.

Kreis, R. (2017) #refugeesnotwelcome: Anti-refugee Discourse on Twitter. *Discourse & Communication*, 11(5): 498–514.

Kress, G. and van Leeuwen, T. (2002) Colour as a Semiotic Mode: Notes for a Grammar of Colour. *Visual Communication*, 1(3): 343–68.

Kress, G. and van Leeuwen, T. (2006). *Reading Images: The Grammar of Visual Design* (2nd edn). London: Routledge.

Krstić, A., Parry, K. and Aiello, G. (2017) Visualising the Politics of Appearance in Times of Democratisation: An Analysis of the 2010 Belgrade Pride Parade Television Coverage. *European Journal of Cultural Studies*, Online First.

Kuang, C. (2014) Why Airbnb's Redesign is All about People, not Places. *Wired*. Available at: www.wired.com/2014/07/why-airbnbs-new-branding-strategy-is-all-about-people-not-places/ (accessed 11 March 2019).

Lakoff, G. and Johnson, M. (1980). *Metaphors We Live By*. Chicago: University of Chicago Press.

Lash, S. and Lury, C. (2007) *Global Culture Industry: The Mediation of Things*. Cambridge: Polity.

Lawler, R. (2012) Airbnb Launches Neighborhoods, Providing the Definitive Travel Guide for Local Neighborhoods. *Tech Crunch*. Available at: https://techcrunch.com/2012/11/13/airbnb-launches-neighborhoods-providing-the-definitive-travel-guide-for-its-guests/ (accessed 10 January 2010).

Lecher, C. (2017) Here are the Russia-linked Facebook Ads Released by Congress. *The Verge*. Available at: www.theverge.com/2017/11/1/16593346/house-russia-facebook-ads (accessed 15 January 2019).

Ledin, P. and Machin, D. (2018) *Doing Visual Analysis: From Theory to Practice*. London: SAGE.

Lee, B. and Campbell, V. (2016) Looking Out or Turning in? Organizational Ramifications of Online Political Posters on Facebook. *The International Journal of Press/Politics*, 21(3): 313–37.

Lehmuskallio, A., Häkkinen, J. and Seppänen, J. (2018) Photorealistic Computer-generated Images are Difficult to Distinguish from Digital Photographs: A Case Study with Professional Photographers and Photo-editors. *Visual Communication*. Online First.

Lester, P.M. (1995) *Visual Communication: Images with Messages*. Belmont, CA: Wadsworth.

Lester, P.M. and Ross, S.D. (2003) *Images that Injure: Pictorial Stereotypes in the Media* (2nd edn). London: Praeger.

Levin, S. (2017) Did Russia Fake Black Activism on Facebook to Sow Division in the US? *The Guardian*. Available at: www.theguardian.com/technology/2017/sep/30/blacktivist-facebook-account-russia-us-election (accessed 15 January 2019).

Levin, S. (2018) Black Panther's Legacy: Will the Record Breaker Finally Smash Hollywood Bias? *The Guardian*. Available at: www.theguardian.com/film/2018/mar/24/black-panther-highest-grossing-ever-hollywood-diversity-bias (accessed 6 August 2018).

Levin, S., Wong, J.C. and Harding, L. (2016) Facebook Backs Down from 'Napalm Girl' Censorship and Reinstates Photo. *The Guardian*. Available at: www.theguardian.com/technology/2016/sep/09/facebook-reinstates-napalm-girl-photo (accessed 11 January 2019).

Lewis, T. (2008) *Smart Living: Lifestyle Media and Popular Expertise*. New York: Peter Lang.

Liebhart, K. and Bernhardt, P. (2017) Political Storytelling on Instagram: Key Aspects of Alexander Van der Bellen's Successful 2016 Presidential Election Campaign. *Media and Communication*, 5(4): 15–25.

Lim, M. (2013) Framing Bouazizi: 'White lies', Hybrid Network, and Collective/Connective Action in the 2010–11 Tunisian Uprising. *Journalism*, 14(7): 921–41.

Linfield, S. (2010) *The Cruel Radiance: Photography and Political Violence*. Chicago and London: University of Chicago Press.

Lister, M. and Wells, L. (2001) Seeing beyond Belief: Cultural Studies as an Approach to Analysing the Visual. In van Leeuwen, T. and Jewitt, C. (eds) *Handbook of Visual Analysis*. London: SAGE.

Lobinger, K. and Brantner, C. (2015) Likable, Funny or Ridiculous? A Q-sort Study on Audience Perceptions of Visual Portrayals of Politicians. *Visual Communication*, 14(1): 15–40.

Luckman, S. (2013) The Aura of the Analogue in a Digital Age: Women's Crafts, Creative Markets and Home-based Labour after Etsy. *Cultural Studies Review*, 19(1): 249–70.

Lury, C. (2004) *Brands: The Logos of the Global Economy*. London and New York: Routledge.

Lutz, C.A. and Collins, J.L. (1993) *Reading National Geographic*. Chicago: University of Chicago Press.

Machin, D. (2004) Building the World's Visual Language: The Increasing Global Importance of Image Banks in Corporate Media. *Visual Communication*, 3(3): 316–36.

Machin, D. and Mayr, A. (2012) *How to do Critical Discourse Analysis*. London: SAGE.

Machin, D. and van Leeuwen, T. (2007) *Global Media Discourse: A Critical Introduction*. London: Routledge.

Madianou, M. (2014) Smartphones as Polymedia. *Journal of Computer-Mediated Communication*, 19(3): 667–80.

Maguire, E. (2015) Self-branding, Hotness, and Girlhood in the Video Blogs of Jenna Marbles. *Biography*, 38(1): 72–86.

Manning, C. (2015) Laverne Cox Opens Up About Her Struggle to Accept Herself in a Transphobic Society. *Cosmopolitan*, 12 September. Available at: www.cosmopolitan.com/style-beauty/news/a46179/laverne-cox-interview-trans-is-beautiful-nyfw/ (accessed 11 March 2019).

Manor, I. and Crilley, R. (2018) Visually Framing the Gaza War of 2014: The Israel Ministry of Foreign Affairs on Twitter. *Media, War & Conflict*, 11(4): 369–91.

Manovich, L. (2017) *Instagram and Contemporary Image*. Available at: http://manovich.net/index.php/projects/instagram-and-contemporary-image (accessed 11 March 2019).

Manzo, K. (2008) Imaging Humanitarianism: NGO Identity and the Iconography of Childhood. *Antipode*, 40(4): 632–57.

Margolis, E. and Pauwels, L. (eds) (2011) *The SAGE Handbook of Visual Research Methods*. London: SAGE.

Martinson, J. (2015) The Virtues of Vice: How *Punk* Magazine was Transformed into Media Giant. *The Guardian*, 1 January. Available at: www.theguardian.com/media/2015/jan/01/virtues-of-vice-magazine-transformed-into-global-giant (accessed 10 December 2018).

Marwick, A. (2013) *Status Update: Celebrity, Publicity and Branding in the Social Media Age*. New Haven, CT: Yale University Press.

Marwick, A. (2015) Instafame: Luxury Selfies in the Attention Economy. *Public Culture*, 27(1): 137–60.

Mattern, S. (2008) Font of a Nation: Creating a National Graphic Identity for Qatar. *Public Culture*, 20(3): 479–96.

Mattoni, A. and Treré, E. (2014) Media Practices, Mediation Processes, and Mediatization in the Study of Social Movements. *Communication Theory*, 24(3): 252–71.

Mayes, S. (2014) Toward a New Documentary Expression. *Aperture Magazine*, Spring. Available at: https://aperture.org/blog/toward-new-documentary-expression/ (accessed 11 January 2019).

Mayes, S. (2017) Beyond the Decisive Moment: 'The Truth isn't Told in a Single Photograph'. *Time*, 15 May. Available at: http://time.com/4775915/everyday-africa/ (accessed 11 January 2019).

McLeod, D.M. and Hertog, J.K. (1992) The Manufacture of 'Public Opinion' by Reporters: Informal Cues for Public Perceptions of Protest Groups. *Discourse & Society*, 3(3): 259–75.

Meeks, L. (2012) Is She "Man Enough"? Women Candidates, Executive Political Offices, and News Coverage. *Journal of Communication*, 62(1): 175–93.

Melucci, A. (1989) *Nomads of the Present: Social Movements and Individual Needs in Contemporary Society*. London: Hutchinson Radius.

Messaris, P. and Abraham, L. (2001) The Role of Images in Framing News Stories. In Reese, S., Gandy, O. and Grant, A. (eds) *Framing Public Life*. Mahwah, NJ and London: Lawrence Erlbaum, pp. 215–26.

Milner, R. (2016) *The World Made Meme: Public Conversations and Participatory Media*. Cambridge, MA: MIT Press.

Miltner, K.M. and Highfield, T. (2017) Never Gonna GIF You Up: Analyzing the Cultural Significance of the Animated GIF. *Social Media + Society*, 3(3): 1–11.

Mirzoeff, N. (2005) *Watching Babylon: The War in Iraq and Global Visual Culture*. New York: Routledge.

Mitchell, W.J.T. (2002) Showing Seeing: A Critique of Visual Culture. *Journal of Visual Culture*, 1(2): 165–81.

Mollerup, P. (1996) *Marks of Excellence: The History and Taxonomy of Trademarks*. London: Phaidon.

Monaco, J. (2004) *How to Read a Film: Movies, Media, and Beyond*. Oxford: Oxford University Press.

Moor, L. (2007) *The Rise of Brands*. Oxford: Berg.

Morris, M. ([1988] 1996) Banality in Cultural Studies. In Storey, J. (ed.) *What is Cultural Studies?* London: Arnold.

Mortensen, M. and Trenz, H.-J. (2016) Media Morality and Visual Icons in the Age of Social Media: Alan Kurdi and the Emergence of an Impromptu Public of Moral Spectatorship. *Javnost – The Public*, 23(4): 343–62.

Moxey, K. (2008) Visual Studies and the Iconic Turn. *Journal of Visual Culture*, 7(2): 131–46.

Müller, L. (2013) *Helvetica: Homage to a Typeface*. Zürich: Lars Müller Publishers.

Müller, M.G. (2007) What is Visual Communication? Past and Future of an Emerging Field of Communication Research. *Studies in Communication Sciences*, 7(2): 7–34.

Nanabhay, M. and Farmanfarmaian, R. (2011) From Spectacle to Spectacular: How Physical Space, Social Media and Mainstream Broadcast Amplified the Public Sphere in Egypt's 'Revolution'. *The Journal of North African Studies*, 16(4): 573–603.

Natarelli, M. and Plapler, R. (2017) *Brand Intimacy: A New Paradigm in Marketing*. Hatherleigh Press.

Neue Logo für Linz 2009 [New logo for Linz 2009] (2006) (Press release, June) Available at: www.ci-portal.de/index.php?id=34&tx_ttnews%5Byear%5D=2007&tx_ttnews%5Bmonth%5D=08&tx_ttnews%5Bday%5D=27&tx_ttnews%5Btt_news%5D=49&cHash=13d82fd1b7 (accessed 13 March 2008).

New 08 brand for Liverpool (2004) (Press release, 9 March) Available at: www.liverpool08.com/News/Archive/2004/SepOct04/New_08_brand_for_Liverpool.asp (accessed 13 March 2008).

Niederer, S. (2018) Networked Images: Visual Methodologies for the Digital Age. Amsterdam: Amsterdam University of Applied Sciences. Available at: https://pure.hva.nl/ws/files/4959407/networked_images_1_.pdf (accessed 11 March 2019).

Nölke, A.-I. (2018) Making Diversity Conform? An Intersectional, Longitudinal Analysis of LGBT-Specific Mainstream Media Advertisements. *Journal of Homosexuality*, 65(2): 224–55.

Nothias, T. (2014) 'Rising', 'Hopeful', 'New': Visualizing Africa in the Age of Globalization. *Visual Communication*, 13(3): 323–39.

O'Leary, A. (2013). The Woman with 1 Billion Clicks, Jenna Marbles. *The New York Times*, 12 April. Available at: www.nytimes.com/2013/04/14/fashion/jenna-marbles.html (accessed 11 March 2019).

O'Neill, D., Savigny, H. and Cann, V. (2016) Women Politicians in the UK Press: Not Seen and Not Heard? *Feminist Media Studies*, 16(2): 293–307.

O'Neill, S.J., Boykoff, M., Niemeyer, S. and Day, S.A. (2013) On the Use of Imagery for Climate Change Engagement. *Global Environmental Change*, 23(2): 413–21.

Olesen, T. (2013) 'We are All Khaled Said': Visual Injustice Symbols in the Egyptian Revolution, 2010–2011. *Research in Social Movements, Conflicts and Change*, 35: 3–25.

Olson, L.C., Finnegan, C.A. and Hope, D.S. (eds) (2008) *Visual Rhetoric: A Reader in Communication and American Culture*. Newbury Park, CA: SAGE.

Osborne, S. (2018) Rohingya Refugee Crisis: Children's Drawings Show Horrific Violence They Suffered in Myanmar. *The Independent*. Available at: www.independent.co.uk/news/world/asia/rohingya-refugee-crisis-myanmar-children-drawings-violence-a8504881.html (accessed 14 September 2018).

Ott, B. and Dickinson, G. (2008) Visual Rhetoric and/as Critical Pedagogy. In Lunsford, A., Wilson, K.H. and Eberly, R.A. (eds) *The SAGE Handbook of Rhetorical Studies*. Thousand Oaks, CA: SAGE, pp. 391–405.

Pan, Z. and Kosicki, G.M. (1993) Framing Analysis: An Approach to News Discourse. *Political Communication*, 10(1): 55–75.

Panofsky, E. ([1955] 1987) *Meaning in the Visual Arts*. Harmondsworth: Penguin.

Papacharissi, Z. (2014) *Affective Publics: Sentiment, Technology, and Politics*. Oxford: Oxford University Press.

Papacharissi, Z. and de Fatima Oliveira, M. (2012) Affective News and Networked Publics: The Rhythms of News Storytelling on #Egypt. *Journal of Communication*, 62(2): 266–82.

Parry, K. (2010a) A Visual Framing Analysis of British Press Photography during the 2006 Israel–Lebanon Conflict. *Media, War & Conflict*, 3(1): 67–85.

Parry, K. (2010b) Media Visualisation of Conflict: Studying News Imagery in 21st Century Wars. *Sociology Compass*, 4(7): 417–29.

Parry, K. (2011) Images of Liberation? Visual Framing, Humanitarianism and British Press Photography during the 2003 Iraq Invasion. *Media, Culture & Society* 33(8): 1185–201.

Parry, K. (2012) The First 'Clean' War? Visually Framing Civilian Casualties in the British Press during the 2003 Iraq Invasion. *Journal of War & Culture Studies*, 5(2): 173–87.

Parry, K. (2015) Visibility and Visualities: 'Ways of Seeing' Politics in the Digital Media Environment. In Coleman, S. and Freelon, D. (eds) *Handbook of Digital Politics*. Cheltenham: Edward Elgar, pp. 417–32.

Parry, K. (2019) #MoreInCommon: Collective Mourning Practices on Twitter and the Iconisation of Jo Cox. In Veneti, A., Lilleker, D. and Jackson, D. (eds) *Visual Political Communication*. Cham: Palgrave Macmillan, pp. 227–46.

Parry, K. (2020) Quantitative Content Analysis of the Visual. In Pauwels, L. and Mannay, D. (eds) *The SAGE Handbook of Visual Research Methods* (2nd edn). London: SAGE.

Parry, K. and Richardson, K. (2011) Political Imagery in the British General Election of 2010: The Curious Case of 'Nick Clegg'. *The British Journal of Politics & International Relations*, 13(4): 474–89.

Pastoureau, M. (2001) *Blue: The History of a Color*. Princeton, NJ: Princeton University Press.

Pauwels, L. (2015a) 'Participatory' Visual Research Revisited: A Critical-constructive Assessment of Epistemological, Methodological and Social Activist Tenets. *Ethnography*, 16(1): 95–117.

Pauwels, L. (2015b) *Reframing Visual Social Science: Towards a More Visual Sociology and Anthropology*. Cambridge: Cambridge University Press.

Peñaloza, L. (1998) Just Doing It: A Visual Ethnographic Study of Spectacular Consumption Behavior at Nike Town. *Consumption, Markets & Culture*, 2(4): 337–400.

Perlmutter, D.D. and Wagner, G.L. (2004) The Anatomy of a Photojournalistic Icon: Marginalization of Dissent in the Selection and Framing of 'a Death in Genoa'. *Visual Communication*, 3(1): 91–108.

Pink, S. (2013) *Doing Visual Ethnography*. London: SAGE.

Pirnia, G. (2013) Regina Spektor: My 'Orange Is the New Black' Theme 'Really Fits'. *Rolling Stone*. Available at: www.rollingstone.com/music/music-news/regina-spektor-my-orange-is-the-new-black-theme-really-fits-68775/ (accessed 5 September 2018).

Pometsey, O. (2018) Black Panther Review: It's a Cinematic Revolution. *GQ*. Available at: www.gq-magazine.co.uk/article/black-panther-review (accessed 7 August 2018).

Poole, E. (2002) *Reporting Islam: Media Representations of British Muslims*. London: I.B. Tauris.

Popp, R.K. and Mendelson, A.L. (2010) 'X'-ing out Enemies: *Time* Magazine, Visual Discourse, and the War in Iraq. *Journalism*, 11(2): 203–21.

Portwood-Stacer, L. (2013) *Lifestyle Politics and Radical Activism*. New York: Bloomsbury.

Potsdam 2010 (2004). Standard Press Release. Available at: www.potsdam2010.com/content/presse/pmitteilungen/documents/Standard_PI_english_Maj20004.pdf (accessed 15 March 2008).

Powell, T.E., Boomgaarden, H.G., De Swert, K. and de Vreese, C. (2018) Framing Fast and Slow: A Dual Processing Account of Multimodal Framing Effects. *Media Psychology*. Online First.

Rampley, M. (2005) Visual Rhetoric. In Rampley, M. (ed.) *Exploring Visual Culture: Definitions, Concepts, Contexts*. Edinburgh: Edinburgh University Press, pp. 133–48.

Raun, T. (2016) *Out Online: Trans Self-Representation and Community Building on YouTube*. New York: Routledge.

Ravelli, L., Adami, E., Boeriis, M., Veloso, F.O. and Wildfeuer, J. (2018) Visual Communication: Mobilizing Perspectives. *Visual Communication*, 17(4): 397–405.

Richards, K. (2016) Put Away the Selfie Stick and Live Like a Local, Urges Airbnb's New Campaign. *Adweek*. Available at: www.adweek.com/brand-marketing/put-away-selfie-stick-and-live-local-urges-airbnbs-new-campaign-170920/ (accessed 12 January 2019).

Richardson, K., Parry, K. and Corner, J. (2012) *Political Culture and Media Genre: Beyond the News*. Basingstoke: Palgrave Macmillan.

Riffe, D., Lacy, S. and Fico, F.G. (2014) *Analyzing Media Messages: Using Quantitative Content Analysis in Research*. London: Lawrence Erlbaum.

Ritchin, F. (2009) *After Photography*. New York: W.W. Norton & Company.

Rodríguez, L. and Dimitrova, D.V. (2011) The Levels of Visual Framing. *Journal of Visual Literacy*, 30(1): 48–65.

Rogers, R. (2013) *Digital Methods*. London: MIT Press.

Rollefson, J.G. (2008) The 'Robot Voodoo Power' Thesis: Afrofuturism and Anti-Anti-Essentialism from Sun Ra to Kool Keith. *Black Music Research Journal*, 28(1): 83–109.

Rose, G. (2016) *Visual Methodologies: An Introduction to Researching with Visual Materials* (4th edn). London: SAGE.

Rose, S. (2014) Helvetica: One Font to Rule Them All. *The Guardian*, 4 March. Available at: www.theguardian.com/artanddesign/2014/mar/04/helvetica-one-font-to-rule-them-all (accessed 10 December 2018).

Ross, J.I. (ed.) (2016) *Routledge Handbook of Graffiti and Street Art*. New York: Routledge.

Rottenberg, C. (2014) The Rise of Neoliberal Feminism. *Cultural Studies*, 28(3): 418–37.

Routledge, P. (1997) The Imagineering of Resistance: Pollok Free State and the Practice of Postmodern Politics. *Transactions of the Institute of British Geographers*, 22(3): 359–76.

Routledge, P. (2012) Sensuous Solidarities: Emotion, Politics and Performance in the Clandestine Insurgent Rebel Clown Army. *Antipode*, 44(2): 428–52.

Rovaniemi 2011 (n.d.) Multilingual Culture and Cultural Diversity. Available at: www.rovaniemi2011.fi/?deptid=16018 (accessed 7 March 2008).

Rovisco, M. and Veneti, A. (2017) Picturing Protest: Visuality, Visibility and the Public Sphere. *Visual Communication*, 16(3): 271–7.

Ruiz, P. (2013) Revealing Power: Masked Protest and the Blank Figure. *Cultural Politics*, 9(3): 263–79.

Ryan, K.B. (1990) The 'Official' Image of Australia. In Zonn, L. (ed.) *Place Images in Media: Portrayal, Experience, and Meaning*. Savage, MD: Rowman & Littlefield, pp. 135–58.

Saeed, A. (2007) Media, Racism and Islamophobia: The Representation of Islam and Muslims in the Media. *Sociology Compass*, 1(2): 443–62.

Saha, A. (2017) *Race and the Cultural Industries*. Cambridge: Polity.

Said, E. (1978) *Orientalism*. New York: Random House.

Said, E. ([1981] 1997) *Covering Islam: How the Media and the Experts Determine How We See the Rest of the World*. London: Vintage.

Salter, J. (2012) Airbnb: The Story Behind the $1.3bn Room-letting Website. *The Daily Telegraph*. Available at: www.telegraph.co.uk/technology/news/9525267/Airbnb-The-story-behind-the-1.3bn-room-letting-website.html (accessed 11 January 2019).

Sandberg, S. (2013) *Lean In: Women, Work and the Will to Lead*. London: W.H. Allen, Ebury Publishing, Random House.

San Filippo, M. (2017) Doing Time: Queer Temporalities and Orange is the New Black. In Barker, C. and Wiatrowski, M. (eds) *The Age of Netflix*. Jefferson, NC: McFarland, pp. 75–97.

Sassatelli, M. (2017) 'Europe in Your Pocket': Narratives of Identity in Euro Iconography. *Journal of Contemporary European Studies*, 25(3): 354–66.

Sawer, M. (2007) Wearing your Politics on your Sleeve: The Role of Political Colours in Social Movements. *Social Movement Studies*, 6(1): 39–56.

Scarles, C. (2004) Mediating Landscapes: The Processes and Practices of Image Construction in Tourist Brochures of Scotland. *Tourist Studies*, 4(1): 43–67.

Schill, D. (2012) The Visual Image and the Political Image: A Review of Visual Communication Research in the Field of Political Communication. *Review of Communication*, 12(2): 118–42.

Schroeder, J.E. and Zwick, D. (2004) Mirrors of Masculinity: Representation and Identity in Advertising Images. *Consumption Markets & Culture*, 7(1): 21–52.

Schultz, H. and Jones Yang, D. (1997) *Pour your Heart into It: How Starbucks Built a Company One Cup at a Time*. New York: Hyperion.

Sender, K. (2007) Professional Homosexuals: The Politics of Sexual Identification in Gay and Lesbian Media and Marketing. In Barnhurst, K.G. (ed.) *Media Q, Media/Queered: Visibility and its Discontents*. New York: Peter Lang, pp. 89–106.

Senft, T. and Baym, N.K. (2015) What does the Selfie say? Investigating a Global Phenomenon. Introduction. *International Journal of Communication*, 9: 1588–606.

Serafinelli, E. (2018) *Digital Life on Instagram: New Social Communication of Photography*. Bingley: Emerald.

Shaheen, J. (2001) *Reel Bad Arabs: How Hollywood Vilifies a People*. Northampton, MA: Olive Branch Press.

Sharma, S. (2012) Black Twitter? Racial Hashtags, Networks and Contagion. *New Formations: A Journal of Culture/Theory/Politics*, 78(1): 46–64.

Shifman, L. (2014) *Memes in Digital Culture*. Cambridge, MA: MIT Press.

Shore, C. (2017) "100% Pure New Zealand": National Branding and the Paradoxes of Scale. In Hannerz, U. and Gingrich, A. (eds) *Small Countries: Structures and Sensibilities*. Philadelphia, PA: University of Pennsylvania Press, pp. 47–66.

Shorty (2013) O2 Be More Dog, Winner in Telecom: About this Entry. *Shorty Awards*. Available at: https://shortyawards.com/6th/o2-be-more-dog (accessed 6 August 2018).

Šisler, V. (2008) Digital Arabs: Representation in Video Games. *European Journal of Cultural Studies*, 11(2): 203–20.

Sivulka, J. (2012) *Soap, Sex, and Cigarettes: A Cultural History of American Advertising* (2nd edn). Boston, MA: Wadsworth, Cengage Learning.

Skey, M. (2011) *National Belonging and Everyday Life: The Significance of Nationhood in an Uncertain World*. Basingstoke: Palgrave Macmillan.

Smets, K. and Bozdağ, Ç. (2018) Editorial Introduction. Representations of Immigrants and Refugees: News Coverage, Public Opinion and Media Literacy. *Communications*, 43(3): 293–9.

Smith-Prei, C. and Stehle, M. (2016) *Awkward Politics: Technologies of Popfeminist Activism*. Montreal: McGill-Queen's University Press.

Sontag, S. (1979) *On Photography*. Harmondsworth: Penguin.

Sontag, S. (2003) *Regarding the Pain of Others*. London: Penguin.

Stanczak, G.C. (2007) *Visual Research Methods: Image, Society and Representation*. London: SAGE.

Stanitzek, G. (2009) Reading the Title Sequence (Vorspann, Générique). *Cinema Journal*, 48(4): 44–58.

Stelter, B. (2013) Daredevil Media Outlet Behind Rodman's Trip. *The New York Times*, 3 March. Available at: www.nytimes.com/2013/03/04/business/media/dennis-rodman-in-north-korea-with-vice-media-as-ringleader.html (accessed 11 March 2019).

Steven, R. (2016) Gretel's 'Unbranded' Branding for Vice TV Channel, Viceland. *Creative Review*, 3 March. Available at: www.creativereview.co.uk/gretels-unbranded-branding-for-vice-tv-channel-viceland/ (accessed 10 December 2018).

Sturken, M. and Cartwright, L. (2017) *Practices of Looking: An Introduction to Visual Culture* (3rd edn). Oxford: Oxford University Press.

Tagg, J. (1988) *The Burden of Representation: Essays on Photographies and Histories*. London: Macmillan.

Tait, S. (2008) Pornographies of Violence? Internet Spectatorship on Body Horror. *Critical Studies in Media Communication*, 25(1): 91–111.

The Everyday Projects (n.d.). About. Available at: https://www.everydayprojects.org/about (accessed 10 July 2019).

Thompson, J.B. (1995) *The Media and Modernity: A Social Theory of the Media*. Cambridge: Polity.

Thompson, J.B. (2005) The New Visibility. *Theory, Culture & Society*, 22(6): 31–51.

Thorson, K., Driscoll, K., Ekdale, B., Edgerly, S., Thompson, L.G., Schrock, A., Swartz, L., Vraga, E.K. and Wells, C. (2013) YouTube, Twitter and the Occupy Movement. *Information, Communication & Society*, 16(3): 421–51.

Thumim, N. (2012) *Self-Representation and Digital Culture*. Basingstoke: Palgrave Macmillan.

Thurlow, C. and Aiello, G. (2007) National Pride, Global Capital: A Social Semiotic Analysis of Transnational Visual Branding in the Airline Industry. *Visual Communication*, 6(3): 305–44.

Thurlow, C. and Jaworski, A. (2013) Silence is Golden: The 'Anti-communicational' Linguascaping of Super-elite Mobility. In Jaworski, A. and Thurlow, C. (eds) *Semiotic Landscapes: Language, Image, Space*. London: Continuum, pp. 187–218.

Tiidenberg, K. (2018) *Selfies, Why We Love (and Hate) Them*. Bingley: Emerald.

Tilly, C. (2008) *Contentious Performances*. Cambridge: Cambridge University Press.

Tufte, E.R. (1997) *Visual Explanations: Images and Quantities, Evidence and Narrative*. Cheshire, CT: Graphics Press.

van Dijk, T. (2013) CDA is NOT a Method of Critical Discourse Analysis. *l'Associació d'Estudis sobre Discurs i Societat*. Available at: www.edisoportal.org/debate/115-cda-not-method-critical-discourse-analysis (accessed 10 January 2019).

van Leeuwen, T. (2001) What is Authenticity? *Discourse Studies*, 3(4): 392–7.

van Leeuwen, T. (2005) *Introducing Social Semiotics*. Abingdon: Routledge.

van Leeuwen, T. (2006) Towards a Semiotics of Typography. *Information Design Journal + Document Design*, 14(2): 139–55.

van Leeuwen, T. and Jewitt, C. (2001) *Handbook of Visual Analysis*. London: SAGE.

van Zoonen, L. (2005) *Entertaining the Citizen: When Politics and Popular Culture Converge*. Lanham, MD: Rowman & Littlefield.

van Zoonen, L. (2006) The Personal, the Political and the Popular: A Woman's Guide to Celebrity Politics. *European Journal of Cultural Studies*, 9(3): 287–301.

Veneti, A., Jackson, D. and Lilleker, D. (eds) (2019) *Visual Political Communication*. Cham: Palgrave Macmillan.

Virgin (n.d.). Our Brand. Available at: www.virgin.com/virgingroup/content/our-brand-0 (accessed 10 January 2019).

Vis, F. and Goriunova, O. (2015) The Iconic Image on Social Media: A Rapid Research Response to the Death of Aylan Kurdi. *Visual Social Media Lab*. Available at: http://visualsocialmedialab.org/projects/the-iconic-image-on-social-media (accessed 14 January 2019).

Vivienne, S. (2016) *Digital Identity and Everyday Activism: Sharing Private Stories with Networked Publics*. London: Palgrave Macmillan.

Vivienne, S. (2017) 'I Will not Hate Myself Because You Cannot Accept Me': Problematizing Empowerment and Gender-diverse Selfies. *Popular Communication*, 15(2): 126–40.

von Borries, F. (2004) *Who's Afraid of Niketown? Nike-Urbanism, Branding, and the City of Tomorrow*. Rotterdam: Episode Publishers.

Wagner, A., Trimble, L., Sampert, S. and Gerrits, B. (2017) Gender, Competitiveness, and Candidate Visibility in Newspaper Coverage of Canadian Party Leadership Contests. *The International Journal of Press/Politics*, 22(4): 471–89.

Wainwright, O. (2014) Is it Balls, Vagina, or Both? Airbnb Logo Sparks Wave of Internet Parodies. *The Guardian*. Available at: www.theguardian.com/artanddesign/architecture-design-blog/2014/jul/18/balls-vagina-both-airbnb-logo-internet-parodies (accessed 14 January 2019).

Waites, R. (2011) V for Vendetta Masks: Who's Behind Them? *BBC News online*. Available at: www.bbc.co.uk/news/magazine-15359735 (accessed 10 January 2019).

Waldau, P. (2013) *Animal Studies: An Introduction*. Oxford: Oxford University Press.

Walker-Rettberg, J. (2014) *Seeing Ourselves Through Technology: How We Use Selfies, Blogs and Wearable Devices to See and Shape Ourselves*. Basingstoke: Palgrave Macmillan.

Wardle, C. and Derakhshan, H. (2017) Information Disorder: Toward an Interdisciplinary Framework for Research and Policy Making. *Council of Europe*. Available at: https://edoc.coe.int/en/media/7495-information-disorder-toward-an-interdisciplinary-framework-for-research-and-policy-making.html (accessed 14 January 2019).

Warrington, S. and Crombie, J. (2017) The People in the Pictures: Vital Perspectives on Save the Children's Image Making. London: Save the Children. Available at: https://resourcecentre.savethechildren.net/library/people-pictures-vital-perspectives-save-childrens-image-making (accessed 14 March 2019).

Weber, M. (1978) *Economy and Society: An Outline of Interpretive Sociology*. Berkeley, CA: University of California Press.

Weintraub, D. (2009) Everything You Wanted to Know, but were Powerless to Ask. In Kenney, K. (ed.) *Visual Communication Research Designs*. New York and London: Routledge, pp. 198–222.

Wernick, A. (1991) *Promotional Culture: Advertising, Ideology, and Symbolic Expression*. London: SAGE.

Wilde, A. (2018) *Film, Comedy, and Disability: Understanding Humour and Genre in Cinematic Constructions of Impairment and Disability*. London: Routledge.

Williams, R. (1976) *Keywords: A Vocabulary of Culture and Society*. London: Croom Helm.

Williams, R. (2005) Advertising: The Magic System. In Williams, R. (ed.) *Culture and Materialism*. London: Verso, pp. 170–95.

Williamson, J. (1978) *Decoding Advertisements: Ideology and Meaning in Advertising*. London: Marion Boyers.

Wodak, R. and Meyer, M. (2016) Critical Discourse Studies: History, Agenda, Theory and Methodology. In Wodak, R. and Meyer, M. (eds) *Methods in Critical Discourse Studies* (3rd edn). London: SAGE, pp. 1–22.

Wolfsfeld, G., Segev, E. and Sheafer, T. (2013) Social Media and the Arab Spring: Politics Comes First. *The International Journal of Press/Politics*, 18(2): 115–37.

Womack, Y. (2013) *Afrofuturism: The World of Black Sci-Fi and Fantasy Culture*. Chicago: Lawrence Hill Books.

Woodward, K. (1997) *Identity and Difference*. London: SAGE.

Woodward, K. (2002) *Understanding Identity*. New York: Oxford University Press.

Wotanis, L. and McMillan, L. (2014) Performing Gender on YouTube. *Feminist Media Studies*, 14(6): 912–28.

Yaeger Media Center (2016) *Everyday Africa – Austin Merrill* (online video) Available at: www.youtube.com/watch?v=U38HCiHd3y8&t=12s (accessed 25 September 2018).

Zappavigna, M. (2016) Social Media Photography: Construing Subjectivity in Instagram Images. *Visual Communication*, 15(3): 271–92.

Zarzycka, M. (2016) Save the Child: Photographed Faces and Affective Transactions in NGO Child Sponsoring Programs. *European Journal of Women's Studies*, 23(1): 28–42.

Zelizer, B. (1998) *Remembering to Forget: Holocaust Memory Through the Camera's Eye*. Chicago and London: University of Chicago Press.

Zelizer, B. (2004) When War is Reduced to a Photograph. In Allan, S. and Zelizer, B. (eds) *Reporting War: Journalism in Wartime*. London: Routledge, pp. 115–35.

Zhao, S. and Zappavigna, M. (2018) Beyond the Self: Intersubjectivity and the Social Semiotic Interpretation of the Selfie. *New Media & Society*, 20(5): 1735–54.

Zimmer, R. (2003) Abstraction in Art with Implications for Perception. *Philosophical Transactions – Royal Society of London*, 358(1435): 1285–91.

Zuboff, S. (2019) *The Age of Surveillance Capitalism: The Fight for a Human Future at the New Frontier of Power*. London: Profile Books.

INDEX

Page numbers in *italics* refer to figures; those in **bold** indicate tables.

abstraction: European Capital of Culture (case study) 80–2
academic journals 7
Acquah, N.K. 68
Adami, E. 216, 218–19
Adorno, T. and Horkheimer, M. 189
advertising and promotional culture 186–8, 206–7
   animals in 191–2
      O2's 'Be More Dog' (case study) 192–9
   film trailer as promotional text: *Black Panther* (case study) 199–206
   semiotic analysis 188–9
   visual content analysis 189–90
aesthetics
   of Instagram 43–4
   and lifestyle 211–12
   multimodal, of Vice.com 216–20
Afrofuturism 201, 203–4, 205–6
age/older people, media portrayals of 91
Airbnb, rebranding (case study) 239–47
Anderson, B. 62–3, 86
animals *see under* advertising and promotional culture
Apple logo 238–9, 250, 251–2
Arab Spring 138
   Egyptian uprising 142
architecture: European Capital of Culture (case study) 80
archive imagery and intertextuality 127–8
Association for Education in Journalism and Mass Communication (AEJMC) 7
authenticity
   feminist stock photography in Lean In Collection by Getty Images (case study) 223–4
   rebranding Airbnb (case study) 239–47

Bachman, J. 151, *152*, 153, 154, 156
Barthes, R. 8, 9, 21, 24–5, 28, 29, 32, 126–7, 180, 188
Benjamin, W. 9
Berelson, B. 22, 23

Berger, J. 8, 9, 86–7, 191
Birdsell, D.S. and Groake, L. 30
Black Lives Matter movement (case study) 149–57
Black Panther: film trailer as promotional text (case study) 199–206
Black Panthers and KKK ('Blacktivist' account, Facebook) 127–8
Black Twitter identities 88
bombardment of images 3
brand continuity 237
'brand image' 235, 236
brands/branding
   beyond advertising 234–6
   dimensions of 236–7
   global, of activism 148–9
   national 66–7
   politics of authenticity in rebranding Airbnb (case study) 239–47
   power of logos 237–9
   reception-led visual analysis of digital media and technology brands (case study) 247–53
   self- 38–40, 50
   visual power of 254
British Airways 66
Butler, J. 38

Capa, R. 64, 65
capitalism 188–9
Cartier-Bresson, H. 64–5
celebrity culture 211–12
Chesney, R. and Citron, D. 260–1
children's drawings: Save the Children's 'Remember Rohingya' advert campaign (case study) 102–6
Chouliaraki, L. and Stolić, T. 100, 105–6
'circuit of culture' model 20–1
Clarke, G. 97–8
'classic' selfies 42, 52, 53
Clinton, Hillary 41, 129–30
'cloaked websites' 126
collective identities 62–84

European Capital of Culture (case study) 75–83
Everyday Africa photography project (case study) 68–75
iconic and humanist photography 63–5
imagined communities 62–3, 86
national and post-national identities for global stage 65–7
colour
  in advertising 196–7
  in logos 78, 251, 252
  and texture, as semiotic resources 143, 144, 146–8
  Vice.com 219–20
commodities 261–2
community brand, Airbnb as 242–3
compositional meaning 28, 240
compositional modality 20
conflict and war
  burden of war image 182–3
  role of photojournalistic images 160–1
    objective record of pain of others 161–3
    visual framing of ISIS propaganda in UK press (case study) 167–74
    visual news frames 163–4
  visual icons 164–7
    Alan Kurdi 166–7
    moral spectatorship on social media 166–7, 181–2
    Omran Daqneesh: boy in the ambulance (case study) 175–82
    and viral icons 165–6
consumer culture
  and capitalism 188–9
  lifestyles and media 210–11
content analysis 22–4, 32
  see also visual content analysis
Cox, Laverne 50–8, 99
critical analysis 10–11
critical discourse analysis 26–7, 216
  see also visual discourse analysis
critical observer of social media 167, 181
cultural associations of brands 236
culture, concept and meanings of 5

Daily Mail 119, *120*, 121–2, **170**
Daily Mirror 169, **170**, 171, 173
Daniels, J. 126
Daqneesh, Omran: boy in the ambulance (case study) 175–82
data journalism 260
Davis, A. 186–7
'decisive moment' 64, 75
DeLuca, K. and Peeples, J. 139, 142–3, 150, 151
denotation/connotation 24, 25, 28, 179–80

descriptive analysis 10
dialectical approach to advertising 198–9
DiCampo, P. 68, 69–75
difference
  'fixing' 89–90
  see also stereotypes, difference and othering
digital media
  fake content 116–17, 260–1
  reception-led visual analysis of digital media and technology brands (case study) 247–53
  see also Internet (online); social media; *specific sites*
disability, representations of 91–2
discourse analysis
  critical 26–7, 216
  visual 95, 96–9, 101, 102, 106–7, 223–4
disinformation 116–17
Doerr, N. et al. 138–9
Du Gay, P. 20

Echwalu, E. 68
Edwards, J.L. and Winkler, C.K. 30
Egyptian uprising: Arab Spring 142
Eisenstaedt, A. 64
emotional observer of social media 167, 181
empowerment
  feminism (Lean In Collection) 223–6
  Instagram 56–8
Entman, R. 31
ethnic minorities *see* race/ethnic minorities; *entries beginning* Black
European Communication Research and Education Association (ECREA) 6
European Union (EU)
  European Capital of Culture (case study) 75–83
  post-national identity 67
  refugee 'crisis' 88, 99–100
The Europeans (Cartier-Bresson) 65
Evans, Ieshia 149–57
Everyday Africa photography project (case study) 68–75
experiential metaphors 215

Facebook
  'Blacktivist' account 127–8
  logo 252
Fairclough, N. 19, 26–7
'fake news' 116–17, 260–1
fake political advertising on social media (case study) 125–32
'The Family of Man' exhibition, New York Museum of Modern Art 65
feminist stock photography in Lean In Collection by Getty Images (case study) 221–30

figurative meaning and plastic meaning in semiotics 25
'fixing' difference 89–90
Floch, J.-M. 25, 238, 248
font: Vice.com 219
Foucault, M. 26, 38
Franklin, T.E. 64
Frosh, P. 41, 57
functional brand, Airbnb as 241–2

GANs (generative adversarial networks) 260–1
gender
  in adverts 190
  gendered portrayal of politicians 114–15
  performance 38, 44–50
  stereotyping and objectification 91
  see also women
genre rhetoric of film trailers 200, 203–4
Gessen, M. 146
'gesture', selfies as 40, 41, 42
GIFs (Graphics Interchange Format) 115–16
global perspectives
  local-global relationship: European Capital of Culture (case study) 80–3
  national and post-national identities 65–7
Goffman, E. 23, 39, 91, 190
The Guardian 122–4, 128, 169, **170**

Hall, S. 8, 20–1, 22, 25, 58, 62, 63, 91, 192, 193
Hariman, R. and Lucaites, J.L. 63, 64, 165
Harris, J. 260
hashtags
  on Instagram 51–2
  on Twitter 88
Holm, N. 187–8, 189, 191, 198, 206–7
humanist photography 64–5
  Everyday Africa project (case study) 75
  and iconic photography 63–5
humour/comedy
  in advertising 197
  political 115–16
  political cartoons 119, *120*
  stereotypes 91–2, 94

IBM logo 248, 249, 250, 252, 253
iconic photography/visual icons
  Black Lives Matter movement (case study) 150–7
  and Everyday Africa project (case study) 74–5
  and humanist photography 63–5
  see also under conflict and war
iconography and nation branding 66–7
identities *see* collective identities; self in social and digital media
imagined communities 62–3, 86

immigrants
  EU 'crisis' 88, 99–100
  see also refugees
implied selfie 42
The Independent 104, 117, 120, 122, 169–71
The Independent on Sunday 172, 173
inferred selfie 42
Instagram
  aesthetics of 43–4
  Everyday Africa (case study) 68, 70, 72, *73*, 74
  logo 249, 250, 251–2
  trans identity (case study) 50–8
interactive/interpersonal meaning 28, 240
International Communication Association (ICA) 6
International Women's Day marches 148
Internet (online)
  blogs and websites, semiotic analysis of 216
  political campaigning 112–13
  self-made media by activists 141–2
interpretive analysis 10
intersubjectivity 41–2
intertextuality
  Black Lives Matter movement 153–6
  and archival imagery 127–8
  and X-ing out 129
intimate brand, Airbnb as 243–5
ISIS propaganda in UK press (case study) 167–74
Islam and Islamophobia 87–8, 130–1
Ivory Coast
  elevator in government building, Abidjan 69
  Grand-Bassam tourist resort 71–2

jump-cut editing and visual juxtaposition 46–9, 50

Kerman, P. 93, 94, 96
Kernan, L. 200, 205, 206
Khatib, L. 138
King, Jr, Martin Luther 153, *154*
Kohan, J. 93, 94, 96
Koolhaas, R. 67
Kreis, R. 88
Kress, G. and van Leeuwen, T. 8–9, 22, 28, 97, 120, 146, 223, 225
Kurdi, Alan 166–7

layouts
  European Capital of Culture 78–9
  Vice.com 217–18
legitimacy of underrepresented groups 90
Lester, P.M. and Ross, S.D. 90
LGBTQ people
  Orange is the New Black 92
  stereotypes 92
  see also entries beginning trans

lifestyle brand, Airbnb as 245
lifestyle magazines 213
lifestyles
  aesthetics and celebrity 211–12
  consumer culture and media 210–11
  feminist stock photography in Lean In Collection by Getty Images (case study) 221–30
  imagery between text and context 230
  in range of media 212–15
  Vice Media's countercultural lifestyle identity (case study) 215–21
Lister, M. and Wells, L. 9–10
Lobinger, K. and Brantner, C. 114–15, 116
local-global relationship: European Capital of Culture (case study) 80–3
logos
  Airbnb 242, 244–5
  European Capital of Culture 78, 80–2
  power of 237–9
  reception-led visual analysis of digital media and technology brands (case study) 247–53
  Vice.com 215
Lutz, C.A. and Collins, J.L. 23
luxury tourism advertising 214

Machin, D.
  and Mayr, A. 8, 10
  and van Leeuwen, T. 211–12, 213
Magnum photographers 64–5
Maguire, E. 45, 47–8, 49, 50
'mainstream' and 'alternative' media 137–8, 142–3
makeover television shows 212–13
manipulative practices
  and fake content in digital media 116–17, 260–1
  truth-value and 3
Manovich, L. 43–4, 51, 53, 55
Manzo, K. 100–1
Marbles, Jenna on YouTube (case study) 44–50
masks 140
  ISIS propaganda in UK press (case study) 168–74
  Pussy Riot balaclavas 146–8
May, Theresa: on front page during Brexit negotiations (case study) 117–25
Mayes, S. 72, 74, 75
media culture
  definition of 5
  understanding images in 4–5
'media output' analysis 19
memes 115, 116, 117
  familiar conventions and Islamophobia 130–1
  Theresa May 124

Merrill, A. 68, 69–75
Messaris, P. and Abraham, L. 163
methodological approaches 21–32
  choice of 18–21
  reception and production of images 32–5
metonymy: European Capital of Culture (case study) 79–80
mirrored selfies 42, 52, 53
moral spectatorship on social media 166–7, 181–2
Mortensen, M. and Trenz, H.-J. 166–7, 181–2
Mourey, Jenna Nicole see Marbles, Jenna on YouTube (case study)
Müller, M. 6–7
multiple meanings of images 4–5, 248
multiplicity as diversity: European Capital of Culture (case study) 77–9
mundanity: Everyday Africa project (case study) 69–72, 74
Myanmar see Save the Children's 'Remember Rohingya' advert campaign (case study)
'myth', function of 180

Nanabhay, M. and Farmanfarmaian, R. 142
national identities see collective identities
Netflix see Orange is the New Black (OITNB)
news frames see visual framing analysis/news frames
news media 88
news photographs 31, 32
newspapers
  ISIS execution videos (case study) 169–73
  May, Theresa: on front page during Brexit negotiations (case study) 117–25
  and national identity 63
  Save the Children's 'Remember Rohingya' campaign (case study) 102–4
Nike brand 237
9/11 attacks 87, 88
  firefighters raising American flag at Ground Zero (Franklin) 64

O2's 'Be More Dog' (case study) 192–9
objective record of pain of others 161–3
Ocasio-Cortez, Alexandria 258
Occupy movement 141–2
older people, media portrayals of 91
online media see digital media; Internet (online); social media; specific sites
Orange is the New Black (OITNB)
  promotes diversity and difference in title sequence (case study) 93–9
  trans character 50, 99
Orientalism 87–8
other

defining 87–9
see also stereotypes, difference and othering
Oxford Dictionaries 40–1

paradigmatic signs 197
Parry, K. 168
'People are People the World Over' (Capa) 65
performance
   gender 38, 44–50
   identity 38–40
   protest and dissent 138–9
photography 9–10, 161–3
   see also iconic photography/visual icons; Instagram; 'selfies'; stock photography
photojournalism 64–5
   and Everyday Africa project (case study) 75
   see also conflict and war
plastic meaning and figurative meaning in semiotics 25
plurality: Everyday Africa project (case study) 72–5
political advertising on social media 115–17
   fake (case study) 125–32
political communication 110–12
   and imagery in public sphere 111, 132–3
political leadership and media visibility 112–13
political 'optics' 122, 125
political protest *see* protest and activism
politicians, female 113–15, 258
   Theresa May on front page during Brexit negotiations (case study) 117–25
'polysemic'/multiple meanings of images 4–5, 248
Pope, R. and Mendelson, A. 129
portraits
   Orange is the New Black (OITNB) title sequence 96–8
   posed, on Instagram 55–6
post-communist countries 67
post-national identity 67
postcolonial theory 87–8
presented selfie 42
producer-led analysis: Everyday Africa photography project (case study) 68–75
production and reception research methods 32–5
professional photographs 44, 55, 56
promotional communication of national and post-national identities 66, 77–82
promotional culture *see* advertising and promotional culture
propaganda 127–31, 167–74
protest and activism 136–8, 157–8
   Black Lives Matter movement (case study) 149–57
   image politics 138–9

'mainstream' and 'alternative' media 137–8, 142–3
media representation 140–1
Pussy Riot (case study) 143–9
self-made media 141–2
visual markers at events 139–40
public screen 139, 142–3
public sphere
   'amplified public sphere' 142
   political communication and imagery in 111, 132–3
   and public screen 139, 142–3
   role of media in 111
Pussy Riot (case study) 143–9
'Pussyhat Project' 148

Q-methodology 33–4, 248
Qatar 66–7

race/ethnic minorities
   and racism 88–9
   stereotyping 91
   see also entries beginning Black
'Raising the Flag on Iwo Jima' (Rosenthal) 64
Rampley, M. 29
rebranding 238
   Airbnb (case study) 239–47
reception and production research methods 32–5
reception-led visual analysis of digital media and technology brands (case study) 247–53
reflexive spectator of social media 167, 181
refugees
   EU 'crisis' 88, 99–100
   global visual icon 166–7
   Save the Children's 'Remember Rohingya' advert campaign (case study) 100–6
representational meaning 28, 240
rhetoric
   of film trailers 200, 203–5
   see also visual rhetoric
Richardson, K. et al. 5
Riffe, D. et al. 22–3
Rodriguez, L. and Dimitrova, D.V. 32
Rose, G. 6, 9, 10, 19, 20, 22, 34
Rosenthal, J. 64
Routledge, P. 140, 147
Ruiz, P. 140, 146

Said, E. 87–8
San Filippo, M. 93, 99
Save the Children's 'Remember Rohingya' advert campaign (case study) 100–6
Schill, D. 111–12
Schroeder, J.E. and Zwick, D. 190

Second World War and post-war photography 64–5
self in social and digital media 259–60
    Instagram
        aesthetics 43–4
        trans identity (case study) 50–8
    Jenna Marbles on YouTube (case study) 44–50
    and meaning-making 58–9
    performing identity 38–40
    'selfies' 40–3, 52–3
self-branding 38–40, 50
self-made media by activists 141–2
self-representation 38–40, 58
'selfies' 40–3, 52–3, 56
Selfies Research Network 40
semiotics/semiotic analysis 24–6
    advertising 188–9
    O2's 'Be More Dog' (case study) 192–9
    conflict and war images 168, 180
    logos: digital media and technology brands (case study) 247–53
    trans self-representation on Instagram (case study) 51
    Vice.com (case study) 216–20
    see also social semiotics
Senft, T. and Baym, N. 40
'site of the image' 19–20
social class representations 92
social media
    animals in advertising 192
    and digital technologies 3–4, 261–2
    moral spectatorship on 166–7, 181–2
    see also political advertising on social media; self in social and digital media; self-made media by activists; *specific sites*
social modality 20
social movements *see* protest and activism
social semiotics 27–9
    Pussy Riot (case study) 143–9
    rebranding Airbnb (case study) 240–7
Sontag, S. 161–2
Starbucks brand 237, 238
stardom rhetoric of film trailers 200, 204–5
stereotypes, difference and othering
    categories 90–3
    female politicians 113–14
    images enacting social and cultural difference 86–90
    Orange is the New Black promotes diversity and difference in title sequence (case study) 93–100
    Save the Children's 'Remember Rohingya' advert campaign (case study) 100–6
stock photography 213–14

feminist stock photography in Lean In Collection by Getty Images (case study) 221–30
story rhetoric of film trailers 200, 204
Sturgeon, Nicola 121–2
Sturken, M. and Cartwright, L. 22, 25
The Sun 169, **170**, 171
The Sunday Telegraph 171
The Sunday Times 171
symbolism in advertising 193–4, 198
syntagmatic signs 197

technological modality 20
television
    and class identity 92
    makeover shows 212–13
    Netflix *see Orange is the New Black (OITNB)*
    political campaigning 111, 112–13
    title sequences 94
    visual framing 32
texture and colour, as semiotic resources 143, 144, 146–8
Thai Airways 66
Thompson, J.B. 112–13
Thumim, N. 39
Tilly, C. 138–9
Time magazine 129
The Times 169, **170**
tourism 66, 71–2
    luxury tourism advertising 214
trans self-representation on Instagram (case study) 50–8
trans visibility 89–90, 92
Trump, Donald 129, 130, 148
Tumblr blog (Everyday Africa) 68
Twitter
    logo 249, 250, 251, 253
    'Lonely Theresa May' 124
    race and racism 88
    visual icon for #BlackLivesMatter 149–57
typography in logos 81–2

'V-J Day Kiss in Times Square' (Eisenstaedt) 64
van Dijk, T. 19
van Leeuwen, T. 27–8, 217
    Kress, G. and 8–9, 22, 28, 97
    Machin, D. and 211–12, 213
Vice Media's countercultural lifestyle identity (case study) 215–21
viral icons 165–6
Virgin brand 236–7
visibility of underrepresented groups 89–90

visual communication
  mapping studies 5–7
  study approach 8–11
  understanding and defining 3–5
visual content analysis 23
  advertising 189–90
  Lean In Collection (case study) 224–6
visual discourse analysis 95, 96–9, 101, 102, 106–7, 223–4
visual framing analysis/news frames 31–2, 140–1, 163–4
  ISIS propaganda in UK press (case study) 167–74
visual icons *see* conflict and war; iconic photography/visual icons
visual juxtaposition
  and jump-cut editing 46–9, 50
  trans pre- and post-surgery 53–4, 57
visual propaganda 127–31, 167–74
visual rhetoric 29–31, 45–6, 126–7
  analysis of European Capital of Culture (case study) 75–83
visual trope of X-ing out 128–30

Waiswa, S. 68
war
  Second World War and post-war photography 64–5
  see also conflict and war
'ways of seeing' 86–7
Weintraub, D. 26
Wernick, A. 186–7
Williams, R. 5, 21
women 148
  diversity: *Orange is the New Black (OITNB)* title sequence 95–9
  feminist stock photography in Lean In Collection by Getty Images (case study) 221–30
  politicians 113–15, 258
    Theresa May on front page during Brexit negotiations (case study) 117–25
  selfies 40, 41
  see also gender

X-ing out 128–30

YouTube 175, 192
  Jenna Marbles on (case study) 44–50

Zhao, S. and Zappavigna, M. 41–2, 51, 52